Earth Class

Bedside Books
An imprint of American Book Publishing
American Book Publishing
P.O. Box 65624
Salt Lake City, UT 84165
www.american-book.com
Printed in the United States of America on acid-free paper.

Earth Class

Designed by Pete Savage, design@american-book.com

Publisher's Note: *This is a work of fiction. Names, characters, places, and incidents either are the product of the author's imagination, or are used fictitiously, and any resemblance to actual persons, living or dead, events, or locales is entirely coincidental.*

ISBN 1-58982-232-3

Wagner, Larry, Earth Class

Earth Class

By

Larry Wagner

This book is dedicated to

Wendy

Who met me as a young man, outlined my responsibilities to her and our children and absolutely insisted that I accept them without whose insistence I would certainly be much less than I am today.

Preface

About This Book

This is a self-help book. (As all help ultimately is) It is spiritual in nature (As all things ultimately are.) And though it is easy to read, it goes into deep, thought provoking subjects of an inter-personal nature. It does it in a subtle, mostly unobtrusive way, allowing the reader to form his own conclusions about each subject. In fact that is the sole objective of the book. Though the book is short, it will not be a fast read. Each chapter will present a 'kernel' thought, some of them in more detail than others. Then, the reader may want to put the book down and ponder how that subject may pertain to his life. The book will explain why some of us are rich while some of us are poor and what to do about it. It will explain where happiness really comes from and what to do to attract it. It will explain the confusing perception of the 'bad things happen to good people' dilemma. But mostly it will explain that you are the master of your own destiny and will experience the consequences of your own creations. Good or Bad.

If you are deeply committed to a particular religious organization that is static and has no room in it for change or continued growth, but rather believe that all revelations were revealed thousands of years ago and any additional discourse about them would be blasphemy, I encourage you to put this book back on the shelf. Now!

If, on the other hand, you consider yourself a spiritual person, with all that that encompasses, and you are here to grow and enjoy the bounty of life, then this book can be a useful tool.

Contents

Chapter One

The End
Coming Right Up

The old man sat quietly now. He watched the children as they ran up and down the sidewalk chasing each other in a game of tag, but somehow even watching them play just took too much effort. He could feel his strength waning as a great tiredness swept over him. Rest, he thought. He needed rest. He sank into the old porch rocker and, after a deep sigh, drifted off to a gentle sleep.

He hadn't heard the gate open. He always heard it squeak when it was opened. He told everyone he was going to oil those hinges, but he knew he never would. He took comfort in having everyone be first announced by the squeak of that old gate. When his eyes opened, he saw the outline of a man standing in the afternoon sun's shadow. The sight of a stranger on his porch startled him, and he jumped to his feet, his heart pounding.

"What do you want? What are you doing on my porch?" he asked. He was in no mood for a salesman, if that's what this guy was.

"What do you want?" he asked again impatiently.

"Excuse me, Mr. Bender, I didn't mean to startle you. I'm here for our Thursday afternoon appointment." The stranger hesitated, as if waiting for the old man to speak. "Don't you remember you made the appointment to see me? Let me check my calendar. Yes, it's right here," he said, withdrawing a small appointment book from his jacket pocket. "William Franklin Bender, September 9. You are Mr. Bender, aren't you?"

"Yes, but I didn't make any appointment with you. I don't even know who you are. What do you want with me?"

Putting the appointment book back in his jacket and starting to head for the steps, the man said, "Mr. Bender, it's obvious there has been a mistake. I'm terribly sorry to have bothered you. I'll be on my way. Have a good day, sir."

Bill watched as the man started down the porch steps. He wasn't normally this abrupt with people, but the stranger had startled him by walking up on his porch uninvited, especially since he had been dozing. Maybe I was a little harsh on him, he thought. Oh well, it's over now in any case. As he turned to go back to his chair, he noticed that the arthritis in his right knee felt much better. Normally, just turning like that would produce enough pain to cause him to falter. In fact, he felt strangely stronger all over. He must have really needed that nap. The feeling of

Chapter One
The End: Coming Right Up

euphoria was very pleasant. He had no pain anywhere. He took a deep breath and smiled to himself. As he reached for the arm of his chair, he suddenly froze. There, sitting in his rocker, was an old man. He felt a sense of electricity shoot through him. A moment of panic swept him. He blinked his eyes and shook his head. He staggered back against the rail and focused his eyes on the figure. How is this possible? "Oh my God!" he screamed. The old man was *him*. He looked up at the stranger now standing on the sidewalk. The man had stopped and was standing just outside the fence. He looked at Bill, not speaking. Bill looked at the figure in the chair, then back at the man.

Finally the man spoke. "This is what I came to see you about, Mr. Bender. May I come in?"

Bill stood silently. What was happening to him? Was this a dream; was he awake? The man again opened the gate and came up on the porch.

"Why don't you go inside and sit down, Mr. Bender. I think I can clear this all up for you," the man said as he opened the screen door and held his hand out, gesturing toward the open door. Bill searched the man's face for some explanation.

"Please.. inside," the man said again. He was a tall slender man, graceful in his motion, well dressed, tan with a friendly but serious face.

After they were seated in the living room, the man spoke.

"Mr. Bender, my name is Timothy Marshall." Bill said nothing. He sat there as if in a daze.

"I didn't mean to startle you, Bill…may I call you Bill?" Tim continued without waiting for an answer, "but you *did* ask me to come here today."

"What's going on here?" Bill asked, pushing the issue of the appointment aside. "Who are you? Who is that person on the porch?" Bill was starting to recover slightly now.

"I'm sort of ah…well, I guess you could call me your tour guide. I'm here to help you. And as far as that person on the porch goes, well, I didn't look real close, but offhand, Bill, I'd say that was you. Or rather what used to be you, or is at least the body that you were using for the last 82 years," Tim said with a smile in an obvious attempt to lighten things up a bit.

"Oh my God, oh my God," Bill said. "Am I dead?" he asked, looking up into Tim's face.

"How could you be dead? I can't see or hear dead people, and I can surely see and hear you. Besides, dead people can't talk," Tim said.

"Are you an angel?" Bill asked with apprehension in his voice. His brain was struggling now to make some sense of what was happening.

"Not exactly…actually, no! Though I have been called that before. Let me explain," Tim said, and there was a serious tone to his voice. "You're not dead, I'm not an angel, and this is not heaven. I know that's not what you were expecting to hear. That's only because you thought there were only two options: heaven or hell. But listen to me, Bill. In a very real sense you didn't even die. Oh, I know that is what you call it, but you see, in all actuality there is no such thing as death. What you did manage to do was change your level of consciousness, and rather smoothly I'd say," he remarked, looking at the old man slumped in the porch rocker. "I thought that was your decision. Isn't that why you made the appointment with me?"

"Appointment? I didn't make any appointment with you. At least I don't think I did," Bill said, his voice tapering off. "I don't know what you're talking about."

"Well, no matter. It obviously just slipped your mind. It's not that important, really. You'll see," Tim said, trying not to let the conversation bog down, at least for the moment, on an unimportant point

"And 'decision.' You keep saying 'decision,' as if I had decided this," Bill cried, and there was agitation in his voice. "I didn't have anything to do with this. I'm old, a little sick maybe, and weaker than I use to be. That's just the way things go. I didn't want it to happen; no one wants it to happen, it just does."

"Does it now? Tim asked rhetorically with a quizzical look on his face. "Yes, I'm sure you believe that, but don't worry; that's one of your first classes. That is, unless you don't want to go."

"What classes and what do you mean, 'If I don't want to go'?" Bill asked with a piqued interest.

"If you don't want to get on with your life, then stay here. It is your choice. It's always your choice, Bill. I remind you, though, that you called me. You must have thought you wanted to get going, to get on with it."

"Are you saying I don't have to die, that I can return to my body and be alive?" Bill asked

"Well, of course, if you want to. Do you want to? Take a look at that old body. Are you really so eager to jump back into that thing?" Tim gave a shudder as he looked through the window at the wrinkled and withered figure sitting in the rocker. "Look, Bill, you're not making any decision now that's in any way different than the one you make every night of your life just before dropping off to sleep. The only difference now is that you have decided you are not going to expend the energy to go back into a body that has served all the purposes you put to it. If you want to go back, then of course go. You won't remember any of this. You'll just wake up from a nap."

"What do you mean I won't remember any of this? Why not?" Bill asked. For some reason his fear had started to fade. Tim was starting to make him feel somewhat at ease, and the fact that he could go back if he wanted to was comforting, but he would hear a little more about this now. "Why won't I remember any of this if I go back? Wouldn't it be like a dream?" Bill asked.

"Not this time. It has to do with quantum physics," Tim replied

"What?" Bill asked. "What are you talking about?"

"Quantum physics, Bill. It wasn't something you've been too interested in. You'll learn a lot more about it in class, but basically, if you choose to go back to your body, then this conversation will have gone down a time line that will not have existed for you in your old body. Hence, for you this will never have happened. Quantum physics is a very big part of your class curriculum. Concepts such as time and space and what you use them for, as well as your own personal reality, will be discussed in detail. I know you haven't had much interest in science, but you've got a great instructor, and I know you're going to like him."

Well, Bill thought to himself, if I'm not dead, and this isn't heaven, I sure hope it isn't ah…that other place. Bill couldn't bring himself to say the word hell, as if it may somehow become a self-fulfilling prophecy. He wished now that he had been just a little more religious, just in case this guy wasn't who he said he was. Oh, he always believed in God, but he just couldn't buy religion. Now he couldn't help but wonder if he'd made the right choice. Once, many years ago, a local minister had called on him, obviously to get him to come to church. Bill let him know he wasn't interested. Just before the minister left he said, "Well, sir, if one of us is wrong about what is going to happen after we die, I hope for your sake it's me." Were those words coming back to haunt him now? Had he made a mistake? Oh God, forgive me the thought.

"Bill…Bill, snap out of it," Tim said with a note of seriousness in his voice. "Come on now, get ahold of yourself."

Bill looked up and stared deeply into Tim's eyes, then blurted out, "Is this hell?"

Tim laughed, but he knew Bill was serious, and more importantly, afraid. "No, Bill, this is not hell. Just like there is no heaven, there is also no hell. In the days to come this will become much clearer to you. I know you're a little confused right now, but your existence doesn't end with what you call death. Your existence continues, on and on and on. You just have different experiences within your existence, that's all. It's a little complicated. You are going to have to just trust me here for a minute, but basically, physical life is a learning experience designed to help you remember and feel the experience. Much learning is accomplished through direct experience, such as the physical experience you've just completed. However, class assignments and lectures are very helpful, and looking at your schedule, I see you've signed up for some real doozies."

"Classes, you keep talking about classes. What classes? What are you talking about?" Bill asked.

"Whenever a soul passes from Earth consciousness, as you have just done, there is a period of time where there is some confusion. At the moment, you do not remember any of the other lives that you spent on Earth or, for that matter, anywhere else. You're confused about just who you really are and what you're up to. The classes that I am referring to are to help you make sense of what you have just been through and what lies ahead. Some classes will prepare you for another go-around on Earth, hopefully this time with a little more success in some of the areas that you had trouble in. Other classes will introduce you to other dimensions and realities that you would have difficulty understanding right now".

Looking at his watch, Tim said, "Bill, could we continue this later? I don't want to rush you, but we should get going if we are going to go. What's your decision? Coming or staying?" Bill looked out on the porch at the figure in the rocker, then turned and looked deeply into Tim's eyes. He was quiet for a moment, his mind replaying the events of a lifetime. He remembered the days of his youth so long ago. He thought of his wife, Margaret. Oh Maggie, he thought. The times of their lives, the trouble they had along the way, as well as the special moments. He thought of his children, all grown now and doing well. The beauty of it all, he thought. He could almost taste it. Then, for a moment, just a moment, he was sad. He had learned so much here, experienced so much. He had known it would all end some day, but now that day was at hand. Yet in a strange way he knew he was ready. A smile crossed his lips as he took one last look at the familiar sights around him. Then, in the same way that he had embraced every new adventure throughout his life, he turned to Tim and said, "What the hell, lets blow this pop stand"

Chapter Two

An Easy Landing

"There now, Bill, that wasn't so bad, was it?" Tim asked as they entered what appeared to Bill to be a large hall.

"No, I guess not," Bill said, at ease now and looking forward to whatever was coming next. "What is this place?"

"It's Guigan Hall. We're just going to walk through it on the way over to your dorm. I thought you might be interested in seeing something."

"What?" Bill almost shouted, and there was a little apprehension in his voice.

Tim laughed. "Don't worry, Bill, relax. Take a look in this class," he said as he pointed to the classroom door. Bill walked slowly over to the door and looked through the window. After a minute, he turned to Tim and said, "What's going on in there?"

"They're studying, Bill. What does it look like?"

"It looks like total confusion. How can they possibly be studying? Bill asked.

"They are watching films," Tim said.

"I can see that, but there must be, let me see…" Bill said as he took another look through the window. "There's 10,11, no, 13 different films playing in there, and it seems like the volume is turned up full blast on all of them. How can anyone concentrate with all that confusion?"

"Confusion? There is no confusion." Tim said looking closely at Bill. "You see, Bill, no one can see any film other than the one he's watching. In fact each person believes the film he is watching is the only film that's playing."

"How is that possible? My God, what a racket."

Tim went on, "Each person can see the other people, but they have no idea that everyone isn't watching the same thing they're watching. They don't even realize that they are watching a film, and they certainly don't know that they are in a class. They think they are living in, and are a participant in, the film. For them, the film *is* their life and their reality."

"But I don't understand," Bill interjected. "How can there be so many different 'realities'?"

"Well now, little feller," Tim said, summoning up his best John Wayne drawl, "I'll give you a clue. When you were on Earth, didn't everyone there seem to have a different opinion on just about everything?" Without waiting for an answer, Tim went on, "Each person's life seemed vivid and real to him, did it not? You might have asked, how could there be so many different versions of the same thing? Surely, you thought, reality must look the same to everyone. Well, it's exactly the same thing here. Remember, they don't think they're watching a film. They think that they, and everyone else in the room, are experiencing the same thing. Notice I said experiencing. For them the film *is* reality. And by the way, they won't be able to graduate from this very important class until they figure out that they are only watching a film."

"How are they going to be able to do that? Can't anyone help them? Do they get an instructor?"

"Yes, they do. She's right over there in the corner at her desk. Her name is Ann," Tim said, pointing to a matronly silver-haired lady who seemed to be watching one of the students very closely. "But they're going to need more than just an instructor to get out of this class. A lot more!"

"Why doesn't she help them?" Bill asked

"She does help them, Bill, but at the moment they don't know she's there. They can't see her. Remember, they think they're in the movie, and, since she's not in the movie, they have no way of seeing her. They'll eventually figure it out, but they're going to need each other's help. No one, I repeat, no one, can do this alone. They'll have to help each other. There is no other way. Take a look at Andy over there. He's the one who Ann is hovering around; notice how he keeps looking around the room? He's not paying as close attention to the film as the others. He doesn't know what he's looking for; in fact he can't even see anything except what is on the screen at the moment, but his mind is starting to wander from the story line. He's starting to think that there may be more to his existence than just the experience he's having, read that 'movie' he's watching. For some time now, his life/movie just hasn't been as interesting to him as it used to be. Ann is there to help him as soon as he thinks he starts to see or hear something other than the film.

"Heck," Tim went on. "Who knows, Andy may call for an appointment himself soon."

" An appointment?" Bill said with a quizzical look on his face.

"Yes an appointment," Tim said matter-of-factly. "Much like the appointment you were having so much trouble with when we were discussing things back in your living room." Tim smiled and patted Bill on the shoulder. Just then, Ann looked up and waved as if she were an old friend. A warm, wide smile crossed her face as she looked at Bill. Strange, but he had an overwhelming feeling that he had known her before.

"And you call this a beginning class? Whew! I don't think I could ever do that. That must be the hardest thing in the world, er, I mean in the…well, you know what I mean. How can anyone figure out that they're having some sort of ah…ah dream, especially when they don't even know they're asleep?" Bill said as the two men started to walk down the hall.

"Oh, I don't know," Tim said. "*You* did it!"

"I did it?" Bill looked at Tim intently.

"If you hadn't, Bill, you would still be in that class"

"You mean I was in that class once? When? How long ago? I don't remember it," Bill said, searching Tim's face for any additional information.

Chapter Two
An Easy Landing

Tim just laughed, "You will. Relax, this is just your first day."

Bill looked back over his shoulder as the two men neared the door at the end of the building.

"What's that room number? I forgot to look. I'd like to come back here again and just watch," Bill said.

"Yes, I'm sure you would; you seemed to have quite an interest in that class. We don't use room numbers here. All the classes have names," Tim replied

"OK then, Tim, I'll bite. What's the name of that class?" Bill asked

"Earth," Tim replied as if this sort of question was asked every day.

"Earth." Bill was confused; he didn't understand. "As in *Earth*?"

"That's right, Bill. You just graduated from there yourself this morning!" Tim explained.

"I don't understand. You mean I was in that exact room this morning?" Bill said, pointing back down the hall.

"Correct again," Tim said dryly. "As I recall, the last scene playing in your movie was you sitting on the front porch of your house watching some of the neighborhood children at play."

"Wow! Huh?" Bill said under his breath, mostly to himself. "You mean I was in that room just a few minutes ago?"

"Well, not exactly a few minutes ago, but based on your present perception of time, yes, I guess that would be a correct statement. Remember I told you about those physics classes you have coming up? You'll learn all about 'Time' soon enough."

No wonder Ann seemed so familiar and friendly, Bill thought.

"Was Ann, ah…my teacher?" Bill asked, his pace slowing now as he became immersed in thought.

"Yes, she was," Tim replied. "You and Andy were good friends, and Ann was helping you both. The two of you would have long talks about philosophy and religion. You helped each other. In fact, Bill, you wouldn't have graduated had it not been for Andy's help."

"Wait a minute. How were we able to help each other; in fact, how were we even able to talk to each other if we were both watching only our own films?" Bill asked.

"True, you were watching your own films, or rather participating in your own films, but what you didn't see was that each of your films overlapped the other's. You thought you were seeing Andy in your film experience, and he thought he was seeing you in his. It never occurred to you that both of you were actually creating the experience of the film. You both thought that your life, and the story that was unfolding, was separate from each other and everyone else. That sense of separateness, more than anything else in your entire existence in Earth Class, was the cause of most of your fears, most of your feelings of guilt, and the sense of loneliness that sometimes overtook you. Separateness, now if there ever was a hell, that would have to be it," Tim said. There was a far-off look in his eyes, and his voice trailed off as his mind momentarily drifted away in thought.

Chapter Two
An Easy Landing

Composing himself, Tim took in a deep breath and continued, "I know it must seem a bit confusing right now, but don't worry, tomorrow you'll get a debriefing about Earth Class before you start your new classes. Everything will be explained to you then, and you can ask all the questions you like. You've got a good instructor. I know him personally, and he's a great guy. I know you're going to like him. Come on, I'll take you to your room. You've had quite a time—enough, I should think, for one day." Tim opened the door to Bill's room. The room was small and reminded Bill of his old college dorm. There was a twin-sized bed up against the left wall and small utilitarian desk against the right. A single window graced the far wall and the whole room had the look of a hotel just cleaned for the next guest.

The two men talked for a minute, then said good-bye.

"I'll see you in the morning, Bill. Bye," Tim said, closing the door behind him.

Bill sat down on the bed. It had been a day whose beginning he could hardly remember. In a few minutes he was fast asleep.

Chapter Three

Personal Reality
Or We're Just Watchin' the Movies

"Good Morning, class," the instructor said as he wrote his name on the blackboard. As soon as he turned around, Bill saw that it was Tim.

"My name is Tim Marshall and I'm your instructor for Earth Debriefing Class." He looked directly at Bill and winked.

Humm, Bill thought, smiling to himself, no wonder he thinks my instructor is such a great guy.

"I see there are sixteen of us here today," Tim said, clasping his hands together and eyeing the class. "First of all, let me explain that Earth Debriefing Class is the first stop for all of you just returning from an Earth experience. Its purpose is to orient you and make some sense of what has just happened to you, as well as to outline your choices from here on. So let's get started, shall we? By now you all realize that you are no longer in Earth Class. And you thought you had to die to get here," Tim said, and there was a dry humor in his voice.

"I want to congratulate each of you on a fine job. I know you all have many questions, and we will get to them in just a minute, but first let me give you an overview of what has happened to you. Each of you was involved in a story of your own making. You didn't realize you were creating the story; you mistakenly thought 'things' were happening to you that you had little control over. If you had particularly 'bad luck,' you believed that that was just the way things were. Generally, you just tried to make the most of it. It certainly wasn't your fault and surely not of your own making. As you will soon discover, nothing could be further from the truth. Thank goodness you took full credit for all the *good* things that happened to you. That's something, at least. Class, I tell you this; you always have had, as you do now, free will. Nothing, I repeat, nothing, was ever thrust upon you. You may say, 'I certainly didn't ask for all those terrible things that happened to me,' but for reasons that made good sense to you at the time, you did. You each needed the experiences that you had, sometimes if just to learn that you had NO use for them.

"When I congratulated you on graduation from Earth Class, I wasn't patronizing you. I sincerely meant it. Each of you learned the exact lesson *you* needed for graduation. Among the two most important was Sharing and Love. Each of you helped each other. Believe me, there was no other way. You would not accept, nor would you even be able to see, help from outside that class. You had no choice; you had to help each other. In the classes that lie ahead, there will be challenges, but none more difficult than the one you just completed. You had to learn to trust yourself and each other. You had to teach things to each other, things that you didn't even know yourselves. I'll explain more about that later. You were there to remember what you already knew but had forgotten and to bring those remembrances into experiences, and you did. I salute you. So before I get started, I know many of you have questions that you are just dying to get the answers to, pun intended, so who wants to start off?"

Bill raised his hand.

"Yes, Bill, go ahead," Tim said.

"I don't know if this is a good time, but I haven't been able to think of anything but that Earth Class that you showed me last night. OK, I can accept that I was in there, but I'm not sure I understand why everyone was watching a different movie and they each thought that they were all watching the same one. Why didn't they just look over and see what the other people were watching? And would you explain to me again how Andy and I were able to help each other if we were viewing entirely different movies?

"Wow, you are getting right into it, aren't you? Those are good questions, Bill. Let me answer your last question first," Tim said. "It's simple, really. You two were not watching *entirely* different movies. True, there were many, and I must say, significant differences in each of your films. Your individual films had parts that you shared with each other. In other words, there were times when Andy was in your film and times when you were in his. You sort of cooperated with each other. In that way you were able to receive what the other had to give without it seeming out of the ordinary. Everything seemed perfectly normal. It was just part of another ordinary day for each of you. You'll recall that many people came into and out of your life when you were in Earth Class. Some of those people had a profound effect on your life, and you frequently had deep and meaningful relationships with them. Others you only barely recall. Some you associated with for a lifetime and others only briefly. Each, however, had a purpose. Sometimes they were there for you to help, and other times they came to help you. All this went on right under your noses and seemed like just another day in your life. But these interactions were the method whereby you helped each other. Your conversations were your way of sharing. Your desire to help one another without any thought or even any possibility of a return, was a demonstration of your love. Remember, class, I told you that you could not have graduated without mastering those two most important qualities, Sharing and Love. This is actually a very profound and, at the same time, advanced thought. Each of you would do well to spend some time considering what it means to you. It will, of course, mean different things to each of you. But Sharing and Love, no matter at what level, is absolutely essential to your development. If, for example, all you are capable of sharing is a smile, then be certain that you do share it. You cannot advance to any level without someone helping you and you in turn, helping someone else. Sharing and love are the two most effective ways that we can give help. Remember, you get what you give! I know that seems simple, and in fact it is, but many of you have just come from an experience where self and ego played a disproportionate role, and the true path you were intending to head down may have gotten, shall we say, a little sidetracked. You'll hear more about this in the coming days. Meanwhile, does that answer your question, Bill?"

"Huh? Yes, I guess so," Bill replied, still lost in thought and trying to digest what he had just heard.

"All right then, to answer your first question. Why didn't they look over and see their neighbor's film? Because the thought never crossed their minds, and why should it have? They had no reason to think that the film they were watching wasn't the same for everybody. Think back. Didn't you assume when you were with a group of people watching an event, say a football game, that everybody was seeing the same thing? You wouldn't turn to the person next to you and question him as to what he was watching would you? Of course not, because you assumed that he saw the exact same thing you were seeing. But you know what, Bill? He wasn't, and that's the reason why Andy is still in Earth Class and you are not."

"I don't understand," Bill said, and there was a look of genuine puzzlement on his face.

"It's like this, Bill. Andy is still too convinced that his reality is the only reality there is. Do you remember when we were in your living room and I told you that the concept of personal reality would be discussed later? Well, consider this the first discussion. Even though you and Andy had long talks about such matters, Andy, for the most part, thought they were just speculation and interesting philosophical fantasy, and after each conversation he would return to his own reality. Oh, he was a good listener to your 'ramblings,' but the subject matter just didn't touch him as it did you. Even though you weren't entirely sure yourself about many of the things the two of you discussed, you had a genuine desire to help Andy and have him benefit from the things you were discussing. Helping someone without the expectancy of reward or return is love. You did actually teach him, Bill, and he did benefit from it, even though you weren't entirely sure of what you were talking about yourself. That's partly what I meant when I said earlier that you must teach things that you do not even know yourself. More about that later. Even though Andy enjoyed your conversations, he still thought they were mostly philosophical in nature. You, on the other hand, wondered deeply about such things and each conversation moved you a little further down the path. Finally, after many, many attempts you *did* start to see and hear things that weren't in your film. In fact, when you would talk to your friends about those events, well, you can recall how they would look at you with a strange expression on their faces. Isn't it odd? You were the one who was getting glimpses of the truth, and they thought you were the weird one. Anyway, it wasn't long before you were starting to get 'bleed-through.' Soon you were aware that there were other people in the class and they were in fact seeing a different story than you were. In fact, their story/reality was disturbing to you, and the whole situation was becoming quite uncomfortable. That's when you were first able to see Ann. You'll recall her in a dream you had the day before I came to your house. She was the lady with the horse who kept trying to give you a ride all night. I was the English guy that you kept asking for the directions to the staircase."

"Wow, I remember that dream. That was a weird one. At least that's what I thought it was when I woke up that morning," Bill said, his voice starting to taper off. "So that's what was going on!"

"Now you've got it!" Tim said, and there was triumph in his voice. "Remember, Bill, Andy was just as important to your development as you were to his. He provided you with someone to love, someone to be able to share with. Without that, you would still be in Earth Class—watching the movies!

"I'll tell you what Bill. After class you and I'll go down to Earth Class. I think I can clear up some things for you." Looking up at the class, Tim continued, "Anyone else who is interested is certainly welcome to come along."

Sally, sitting in the third row, looked around, then slowly raised her hand, not certain that Tim had finished talking.

"Yes, Sally?" Tim said, acknowledging her.

"You've said 'remember' several times; what were we in Earth Class to remember?" Sally said. There was uncertainty in her voice, as if she thought she was the only one who didn't get it. "What were we supposed to remember?"

"Everything," Tim said, walking toward the blackboard. "You see, Sally, everything is already known by the internal or, if you wish, the greater you. The you that some have called the soul, that unique sense of 'you-ness' that lives behind your eyes, peering out on the World. As you will soon learn, each of you is much more than the ego self that you are so familiar with. One of your jobs in Earth Class was to remember enough of what you internally knew to graduate. You see, Sally, if learning was your only objective in your various trips to Earth Class, then perfect learning would be accomplished when you had learned everything that was already known. Your experience would then be complete when you were little more than a walking encyclopedia packed with the total knowledge of humanity. Things that were not known, though, could never be known. You would be stuck in a static state. There would not, nor could there ever be, any such thing as invention. Let me give you an example: Some time ago in Earth Class, a man named Michael Faraday, who some of you may recall, invented the refrigerator. Now, prior to his inventing the refrigerator, or at least refrigeration, the refrigerator did not exist, so no one could tell him how to do it. It wasn't something he could 'learn.' Where, then, do you suppose he got the idea and the necessary methods to do it? I'll tell you where. He remembered it. It already existed within his greater self as it did within each person in Earth Class. He was just more strongly motivated to remember than the rest of the people. It was already there. He just had to think, focus, and concentrate to bring it through. Once he had accomplished it, he shared it with others, which was the very thing that made it real. There's that 'sharing' word again, class. Sharing is one of the most important ways each of you made your experience in Earth Class real.

Without sharing, any idea or accomplishment has little value, either to you or anyone else. Good ideas are a dime a dozen, but the person who shares them and makes them live for everyone is invaluable. Does that answer your question, Sally?"

"Yes, I think so. Thank you," Sally said politely.

"You're going to hear a lot more about this very subject in the days to come. Sharing is a form of giving, and how many times have you heard in one way or another, It is better to give than to receive? You see, class, it is only what you give away that you actually get. I know that probably seems a little strange to you right now, but consider this: when you give something to someone, who gets it?" Tim said, looking around

"The person you gave it to, of course," Mary Beth piped up.

"That's right, Mary Beth, but the real question here is, *who is that person?*" Tim said, looking at Mary Beth intently, wanting her to stay with his line of questioning.

"What do you mean? I don't know. Whoever it happens to be, I guess," Mary Beth said, not really understanding the question. "Who?"

"Well, lets look at it. Who are the people involved?

"You, and the person you gave it to, I guess," Mary Beth answered.

"Yes, that's correct, but you are going to be surprised about just who it is that you gave it to." Without stopping, Tim went on, "You gave it to the only person you could have given it to, the only person there to receive it, the only possibility you had. You gave it to yourself." Tim waited, looking at the class, letting what he had just said sink in. Listen closely here, class, this is very important. There was really only one person in your reality in Earth Class. You! Oh, I know it didn't seem that way. You thought there were millions, billions of others. But in fact, there was only you. Those 'others' were all different aspects of who you really were. You were not as limited as you presently think you were. This is a very difficult concept to grasp, and we are not going to be able to do it justice here in Earth Debriefing Class, but believe me when I tell you there was only one person in your experience, and you are going to be very surprised when you find out who that person was. That is why when you gave, or shared, with others, you truly benefited yourself. Why? Because…you were the only one there. You were the giver and the receiver. But you had to learn that giving did not diminish you, that you actually gave up nothing. Once you gave up your attachment to something and didn't feel any sense of loss by giving it away, you had learned the true meaning of sharing. You give it away and get it at the same time. What a deal, huh? Don't worry, folks, there will be a lot more about the concept of One in the days to come. I wouldn't leave you up in the air over such an important topic. Remember this everyone, and this is very important: Earth Class was an illusion that each of you made real for purposes and reasons that will soon become clear to you."

The class sat in silence trying to understand what Tim had just said. Finally, Wayne, sitting near the back, blurted out, "You mean Earth Class was like an empty slate, just waiting for us to write our own story on it?"

Chapter Three
Personal Reality: Or We're Just Watchin' the Movies

"That's a pretty good analogy, Wayne," Tim answered. "You each created your own reality. It wasn't thrust upon you. It never is! Giving and sharing was just one way, albeit an important one, that you used to teach yourself how to create what you desired then accept it as your gift to yourself."

"But how can that be?" asked Mary Beth, "I just can't believe that. I mean…" Her voice suddenly tapered off as she became aware of just how disrespectful she must be sounding to her instructor. The rest of the class looked at Mary Beth. She wasn't the only one thinking the same thing, and Tim knew it.

Looking at Mary Beth, Tim said, "You colored yourself in as a scientist in Earth Class, didn't you, Mary Beth?" "Colored myself in?" Mary Beth questioned.

"All right then, you were a scientist, were you not?" Tim queried

"Yes, a physicist," Mary Beth said

"OK, then, perhaps you will be so kind as to explain to the rest of the class the difference between a virtual and a real particle," Tim asked, and folded his arms across his chest

"Well, that's pretty new stuff. It's not generally understood by lay people. In fact, not all physicists have a good grasp on it," Mary Beth said.

"But you understand it, don't you, Mary Beth?" Tim asked with the assurance in his voice one has when one already knows the answer.

"Yes," she said

"Well then?" Tim said

"Simply put, we have recently discovered that there are certain particles in the universe that appear very quickly, out of nowhere, travel a very short distance, then just as quickly disappear. Before they disappear, they are called 'virtual' particles. They are just like real particles, except they are living on borrowed time, or rather, I should say, borrowed energy. When they 'pay the energy back,' and they must pay it back, they disappear. But for a while, a very, very short while, they are real. But the thing that makes it so strange," Mary Beth said, looking at the class now, speaking very comfortably, as if she was giving one of her many lectures, "is that every once in a while, an outside source of energy, either electrical or magnetic, interferes and pays the debt for the virtual particle and it doesn't have to disappear. It stays. It converts from a virtual particle to a real particle and continues its existence just as if it had been a real particle all along." Mary Beth, suddenly feeling somewhat self-conscious, looked around the room and at Tim, then quickly sat down.

"Thank you, Mary Beth. Very good, very good, "Tim said, applauding.

"Do you see any similarities?" Without waiting for an answer, Tim went on. "Just like Mary Beth's virtual particles that spring into reality, so did all the life experiences you had in Earth Class spring into your reality. Remember, I told you that reality could more accurately be stated as your personality reality. Virtual particles are only one of the ways in which reality is created. Like Wayne's 'slate,' reality was just waiting for you to command something. Soon, you will be learning about even more powerful and, dare I say, fantastic ways that you will be able to take control of creation on your next trip to Earth Class. Remember, everything, in one way or another, springs from you. There is no other place it can come from in Earth Class except from you.

"Some of you will remember there was a personality in Earth Class some time back who produced fish and bread to feed hundreds of people. It was reported by the observers of the day that he produced them out of thin air. Do you think he knew about virtual and real particles?" Tim asked, with a bit of his humor shining through.

Tim looked down at his watch. "OK, folks, I think that's enough for today. Let's head down to the Earth Class room, shall we, and see if we can't clear up a few things for Bill." After everyone had gathered around the door to the Earth Class room, Tim motioned and said, "Step up here, Bill."

Bill walked up to the door and looked in the window. After a minute, he stepped back and said to no one in particular, "Yes, that's what I saw yesterday."

"Good," Tim said. "Now, as I recall, you wanted to know why they couldn't tell that they were each watching a different movie. Well, let's just step inside, shall we?" Opening the door, he said, "Oh, don't worry, you won't disturb them. You could run a train through here and it wouldn't disturb them. Their only reality is what they are seeing on their screens. If it wasn't that way, they would have been out of here long ago."

"What do you mean, Tim? Why would they have been out long ago?" Bill asked.

"Well, Bill, what are we doing in here?" Without waiting for an answer, Tim went on. "I know, I know you wanted to figure out why they can't see that they're all watching different movies, but there's more to it than that, and, unless I miss my guess, Bill, it won't be long before you are going to try to help Andy."

"Sure, why not," Bill said matter-of-factly.

"Go ahead, Bill, help him," Tim challenged. "No, seriously, go help him; it's all right."

Bill slowly approached Andy. He stopped and looked back at Tim.

"Go ahead," Tim said.

Bill moved up close to Andy and slowly sat down beside him. After a moment, he spoke.

"Hi, Andy." Andy paid no attention. Bill looked up at Tim, then spoke again, this time a little louder.

"Andy, it's me, Bill." Still there was no response. Bill looked to Tim again. Tim remained motionless, just staring back at Bill. This time Bill put his hand on Andy's shoulder.

"Andy, Andy!" he said, raising his voice louder and louder. "It's me, Bill. Can't you hear me?"

"No, Bill, he can't," Tim finally said.

Bill stood up and placed himself directly in front of Andy and shook him by the shoulders. Still there was no response. Suddenly, Ann was at Bill's side.

"Oh, thank you, Bill, thank you," Ann said.

"For what?" Bill said, stepping back.

"Didn't you see that?" Ann said excitedly.

Chapter Three
Personal Reality: Or We're Just Watchin' the Movies

"See what? I didn't see anything," Bill answered with a tone of confusion in his voice.

"Andy shifted when you stood in front of him. Didn't you see that? He saw you, Bill. In his mind, you registered as a thought. He thinks he just had a thought of his old friend Bill. But you caused it just now, this second. Your relationship with Andy was strong, and for a split second you brought him to the edge. Thank you, Bill. Thank you."

"Class, lets all leave the room!" Tim said and there was a tone of seriousness in his voice. "Ann has some serious work to do and we don't need to distract her."

After everyone had assembled outside the room, Tim spoke. "Wow, well that certainly wasn't what I expected to happen, but it was fantastic. It just goes to show you ANYTHING can happen"

"What did happen?" Bill asked, turning away from the window. "I don't understand.

"Bill, I took everyone in there to show you that people in Earth Class are so strongly invested in their illusion that nothing can rouse them from it, not even us tramping through their class. Instead, you all just witnessed the power of LOVE. Love will be one of the most important classes you will take here, because love is the most powerful force in the universe—and that is an understatement. As with any such power, it is important you understand it. I say you witnessed it, because even though it was unexpected, Bill's love and his desire to share with his friend, Andy, was so strong it reached through the illusion, and the souls of two friends were united through love. The reason Ann was so excited was because she knows that it is now only a matter of time; Andy's transition is assured. Ann is an expert in coaxing the dormant powers that we all have to the surface. I wanted to get you out of her way. She'll be pretty busy for a while attending to Andy's birth. Yes, class, birth is a good name for it. Unless I miss my guess, we'll all be seeing Andy real soon. I suppose I should explain though, since I didn't expect this to happen, that while we will be seeing Andy in what will seem like a few hours to us, it may be many years to Andy in his realm. But the real lesson here, even though it did come a little out of order is the incredible power of love."

Chapter Four

Everybody's Talkin' 'bout Heaven

The next morning, Bill was the first one to class. The events of the previous afternoon were still fresh in his mind. Even so, he could hardly wait to see what this day would bring. He was now ingesting information like a streamlined computer, and he wanted more. Within a few minutes, the rest of the class had arrived and taken their seats. Tim stood up from his desk.

"Good morning, class." I trust you all had a good rest?"

"Good morning, Tim," several of the class members said almost in unison.

Even though he was sitting quietly, Tim could see that Bill was literally bursting to ask a question.

"Good morning, Bill," Tim said, looking right at Bill as he offered the additional greeting. "Unless I miss my guess, I'll bet you've got a question."

Bill straightened up in his seat. “Well, yes, actually I’ve got lots of questions.”

“Why did I suspect that?” Tim said, with a wry look. “Come on then, let’s have it.”

“I was just thinking…I mean, you know, where are our friends and relatives? Aren’t they supposed to be here? I thought when you died, err…you know, graduated, you could see your family and friends that ah…you know, *graduated* before you did.

“Really?” Tim said with humor in his voice “Who told you that?”

“Ah…well, I don’t know, I’ve just always heard that,” Bill stammered.

“Of course you did. But remember, Bill, it was your creation, so you can hear anything you want to hear,” Tim said. “But yes, you can see your friends. Several of them are in various classes here right now. I’m sure they are looking forward to seeing you as well.”

Judy Lee raised her hand. “Yes, Judy?” Tim said, observing a look of distress on her face.

“Ah…in my illusion…dream…Earth Class,” Judy said as she tried to find the right word, not yet comfortable with the conditions she found herself in, “my husband and I both ah…well, we were both in the same car and ah…we both ah…*graduated* at the same time. I can’t seem to find him. Can you tell me where he is?” She struggled to hold back tears, not sure she wanted to hear the answer.

“Sure, Judy, he’s in heaven,” Tim replied matter-of-factly.

Chapter Four
Everybody's Talkin' 'bout Heaven

Bill couldn't help but notice that Tim seemed rather flippant about an issue that was obviously of great concern to Judy. Judy put her hands up to her cheeks and gasped.

"But I don't understand. I thought, I mean…," Judy stammered.

Tim sensed he had pushed this as far he could. He often used shock treatment as an aid in getting his students to wake up, but right now Judy needed his help. He walked up to Judy's desk, took both of her hands in his, and said, "Judy, what are we trying to learn here? That we made our own reality, right? Earth Debriefing Class is just to help you wake up from a long dream. A dream about things that you made real and then believed. You see, Judy, in Earth Class you first project an image, then you see it, then you believe it without realizing that you were the one who projected it in the first place. Remember, you see, hear, and believe exactly what you expect to see and hear and believe, no more, no less. The same, of course, is true for Jerry."

Judy wondered for just a moment how Tim knew her husband's name was Jerry, but dismissed the thought, confident that Tim would have a logical answer, and, at the moment, she was more interested in knowing where to find Jerry. "Look, Judy, when the two of you were in Earth Class, what were your religious beliefs?"

"We were Christians. We went to church," Judy said

"Which one of you was the *most* religious? Tim asked.

"We both believed in God," answered Judy.

Tim said nothing but continued to look at Judy as if he somehow knew she wasn't through.

"Jerry was deeply religious. I believed in God," she repeated, and there was conviction in her voice, "but Jerry, I guess, was more involved with the church than I was. He always said grace before every meal and prayed aloud before going to bed. When the children got older they didn't want to go to church as much and that always bothered him. On more than one occasion he would remind them that they were playing with their eternal souls."

"Where did Jerry think he was going to go after he died, Judy?" Tim asked.

"Oh, there never was any question in his mind. He was going to heaven," Judy said, her voice rising and her eyes opening wide.

"Well then, in light of what we have learned this morning about each of us being the creator of our own reality, is there any real surprise as to where Jerry is?" Tim asked, as he studied Judy's eyes for her acceptance of his explanation.

"But I believe in heaven too. I want to go. I want to be with Jerry!" Judy said as she burst into tears.

"Then go, Judy, it's still your reality," Tim said.

"Stop saying that," Judy sobbed. "I don't know what you are saying. I don't know how to get there." Again sensing that he had pushed this as far as was practical, Tim put his arm around Judy in an attempt to comfort her.

"All right now, dry those tears. I'll show you how. Here now," Tim said, handing her a tissue. "But first let me try to explain this to you a little better. Jerry is in what he believes to be heaven. In reality, there is no such actual place, just as there is no actual hell. Your experiences are the result of your expectations. You are a creator—actually a co-creator—but we'll get to that later. It's the same for Jerry. The experiences he is having now are a reflection of his expectations. Think about it, Judy. He couldn't have it any other way. Think of the trauma and harm it would do Jerry if things didn't progress the way he generally expected them to. His whole belief system would be shattered in a way that would cause him great difficulty in being able to trust himself again. Major changes to one's core belief system take 'time,' as you've come to think of it. Heck, Judy, he's probably sitting on some cloud right now playing pinochle with his friends." Tim laughed at his own humor and gave Judy a gentle squeeze around her shoulder. "Don't worry, though, Judy. After about the two-thousandth consecutive game, he'll start looking around for something else to do and even heaven will start to get a little boring. When that happens, he'll be very interested in what you have to tell him. The best thing you can do for him right now is to just give him a little time. Let him come to believe something else on his own, so that he may understand what is really happening. If you went charging in there right now trying to tell him he is wrong, believe me you would do more damage than good. Don't worry, he'll be fine and soon, too, I promise you." Tim smiled at Judy and walked back to the front of the class. "By the way, class, while Jerry thinks he is in heaven, some people think they are going to hell. You can imagine how that must feel, what with all that fire and pitchforks and such. Don't feel sorry for them though, folks, they are actually better off than Jerry."

"How is that possible?" Bill chimed in.

"Well, think about it," Tim answered. "If you were in heaven or hell, which one would you be looking for the door out of first? The one you wanted out of the most, right? Think about it. When people go to what they think of as heaven, I would imagine they wouldn't be in as much of a hurry to leave as the folks who have chosen hell. Follow along with me here, class," Tim said with his usual sense of dry humor. "One would think that the folks in hell would be the first to realize that this whole thing is one big mistake. Meanwhile, the folks in heaven only have boredom as their ally. Some start to get a little suspicious, though, when God never seems to show up for the prayer meetings. But no matter, eventually everyone gets back to class and life goes on. Many of you were involved with the various religions while you were in Earth Class. And while your beliefs varied, depending on which religion or organization you belonged to, most all of you believed in some sort of reward or punishment system. Heaven or hell, if you will. As you each progress through your required and elective classes, you will see that *which* religion you belonged to, or even if you belonged to a religion at all, is not the important question. Any or all of them serve about equally well. The only real issue is that you believe in something. For example, Jerry believes, at the present, that he is in heaven, and you will meet people in your upcoming classes who believed they were in hell. Even this matters very little. The only truly dangerous belief is the one held by the people who do not believe anything. Those people believe that when they die, that's the end—lights out, nothing, zip, nada. Since they believe that nothing will happen to them after death, and since they are the creators of their own reality, nothing will happen. It is very difficult for the various guides and teachers to get to them, because they aren't looking for or expecting anybody. It doesn't matter *what* they believe as much as that they believe something. If they believe the wrong thing it can be corrected, but if they don't believe anything, there is very little that can be done. Notice I said very little that could be done. I didn't say there wasn't anything that could be done. There are very specialized teachers who can help those who have placed themselves in a void."

Chapter Five

How'd He Do That?

Everyone was listening intently, when Tim suddenly looked at his watch and announced that it was getting close to lunchtime and asked if anyone was getting hungry. Strange, Bill thought, he'd been here now for what seemed like the best part of two days and yet he wasn't hungry; in fact, he hadn't even thought about eating. The rest of the class looked around at one another as if to see whom might actually be hungry.

"I don't know how long the others have been here, but I got here yesterday morning," Bill said, then hesitated. "At least I think it was yesterday morning. I haven't had a bite to eat and I'm still not hungry."

"Me neither," said Wayne.

"I've been here four days," announced Sally, "and this is the first time I've even thought about food."

"I have to eat regularly or I'll get sick; I'm diabetic," announced Wayne. His voice tapered off as he realized he hadn't had so much as a bite to eat in nearly three days.

"We don't need food here, do we?" questioned Mary Beth. Tim only smiled and looked around the room as if to see if anyone else had anything to say. All eyes were on Tim. "Very good, everyone. You're all starting to remember," Tim said, and there was victory in his voice.

"I guess I don't have too much problem with us not eating," Mary Beth said. "After all, if we are no longer physical bodies, we shouldn't need food." The scientist in her was starting to come out again. "Oh, excuse me," she said, looking back at Tim somewhat self-consciously. "I didn't mean to interrupt."

"No! No! Not at all. Please, do go on," Tim said, holding out his right hand to Mary Beth.

The rest of the class were shaking their heads in agreement.

"Well, I do have a question," Mary Beth said, turning to look at Tim. "I understand that we don't need to actually eat food, but what will maintain our bodies if we don't give them nourishment?"

"The same thing that maintained them while you were in Earth Class," Tim said smugly.

"That was food. Food maintained our bodies on earth. I mean in Earth Class," Mary Beth challenged.

"Yes, I guess it did—indirectly," Tim said, speaking in the half-riddle manner that the class was now becoming accustomed to.

Chapter Five
How'd He Do That?

"All right, Tim, I'll bite," Bill said, knowing he was going to be the victim of another one of Tim's gotchas. "Why do you say indirectly? The whole process seems pretty simple to me. You chew, swallow, digest, and your body does its thing with the food."

"Let's take a closer look at that," Tim said. Everyone knew this was not going to be as simple as it seemed. "Before your body can *use,* say, a slice of bread, you have to eat it, right?" Everyone nodded in agreement. "Before you can eat it, you have to go to the store and buy it, right?" Again everyone nodded in agreement. "Before you can buy it, the baker had to bake it. Before the baker, the farmer had to harvest the grain; before he could harvest the grain, he had to plant the seed and let it grow. Everybody with me so far?" Tim inquired. "Now, all you really wanted was to eat the bread. Why go through all those other steps?" Without waiting for an answer, Tim went on. "Because you believed they were necessary. Remember what we are trying to learn here: as a creator, what you believe, you make real. Let me explain it another way. When all you wanted to do was eat bread, why mess around with all the intermediate steps? When you planted the wheat, it was just a symbolic act; what you really wanted was bread. You didn't want the little shoot that sprang up. You didn't want the wheat kernels; you didn't even want the flour. You wanted the bread. Now, if you wanted bread, why didn't you just produce bread instead of going through all those unnecessary steps? I'll tell you why," Tim continued without waiting for anyone to answer. "Because you thought it was necessary; you thought you had to. You didn't believe you could produce the bread any other way. And you *always* get exactly what you believe you get.

"Are you telling me that if I had believed that I could produce bread out of…of thin air, it would have happened?" Bill asked.

"That's right, Bill!" Tim replied. "Just the way Mary Beth said it would."

"Me?" Mary Beth sat up straight in her chair. "I didn't say anything about that!"

"Really?" Tim quipped "I'd swear you were the one who gave the lecture on virtual particles!"

"Yes, but that was…" and her voice tapered off as it did when she was in deep thought. Tim was quiet. "Maybe, just maybe…," Mary Beth went on, "the force that pays the energy back for the virtual particle and allows it to become real is not electricity or magnetism. Maybe it's—consciousness!" Mary was deep in thought now, employing that incredible logical mind that had earned her so much respect in the scientific community. Her eyes were wide open, not focusing on anything in particular. " I suppose—yes, that could work. Consciousness could be the force that pays off the energy debt. Consciousness could provide the energy to produce something out of…" She hesitated, sank back into her chair, and looked up directly at Tim. "Nothing!" she said, staring at Tim with a look of discovery on her face. Had she just unraveled one of the great mysteries in physics? Tim only smiled.

"Jesus was the first person in Earth Class to figure that out. Remember the fishes and bread? And now Mary Beth is just the latest. Congratulations, Mary Beth," Tim said.

Chapter Five
How'd He Do That?

"Wait a minute!" Judy interjected. "Jesus was the Son of God. He was able to work miracles. It didn't have anything to do with physics."

"How do *you* explain what he did then? I mean the fish and bread trick?" Tim asked, looking at Judy with his head cocked to one side."

"That was no trick. You make him sound like some sort of a circus magician. He was the Son of God and anything was possible for him!" Judy said, somewhat defiantly.

"All right, Judy. I'll buy that. He was the Son of God. He was a co-creator with God. Would that be fair to say?" Tim asked.

"That's right!" Judy said, and there was confidence in her voice.

"And as a creator he could produce anything he wanted to, right?"

"That's right!" Judy replied again with confidence.

"Then if we were creators, we could also produce anything we wanted to as well, correct?" Tim queried.

"Yes, I suppose, but...but we are not creators. I mean we're creators of our own reality, I understand that, but we are not creators of the world. We're not Sons of God. We can't create the same way he did!"

"Just a minute, Judy," Bill injected. "You said we were creators of our own reality, did you not?"

"Yes," Judy said, turning to look at Bill.

"I mean, that's what we are doing here, isn't it, trying to understand how we create our own reality? Our own reality is the only reality we can see; therefore, it seems like the only reality there is for each of us," Bill said, glancing at Tim.

"Go on, Bill," Tim said, "I think you've got it!"

"Does this mean that we could do the same things Jesus did?" Bill asked. There was an almost frightened look on his face.

"I think so," said Tim. "Isn't that what he meant when he said, '*What I do, you can do and even greater things.'* Wasn't he, after all, a teacher?"

"Yes," Judy said, "but even though he was a teacher, he was still the Son of God."

"And you're not? Is that what you are saying, Judy?" Tim asked.

"Of course I'm not!" Judy answered emphatically, her voice raised. "Not like he was." She blurted out, "He was the only Son of God who could work miracles and talk directly to his Father and stuff no other human could ever do." Judy's voice was on rapid fire now, as if she had to distance herself from what Tim was implying. What if God heard her listening to this! This went against everything she had learned. Her mind was racing now and filled with uncertainty. Tim's voice was low and calm as he resumed speaking.

"Judy, if you can create, and you can make your own reality and all the things around you; if you can transcend death as you have obviously done here, aren't those at least some of the things that are reserved for the Son of God," Tim asked.

"Well, yes," Judy answered somewhat reluctantly.

Chapter Five
How'd He Do That?

"Then maybe you're a *partial* Son of God; at least that's something. But I'll tell you what, Judy, why don't you ask Jesus that question the next time you see him. I see here," Tim said as he looked at the class schedules, "that you've signed up for his class on The Dangers of Judgment. It's a good class; I'm sure you'll like it.

"Well, that's all for now, guys. See you tomorrow," Tim said as he gathered up his briefcase and headed out the door without waiting for any questions from the class. The whole class just sat silently, no one moving.

"Did he say Jesus as in the Son of God Jesus?" Mary Beth asked finally, almost under her voice.

"I…I think so," Wayne muttered.

"He did," said Judy. "And I don't think I like this. In fact, I don't like it at all. What does he mean?"

The class slowly got up and, one by one, filed out of the room. They had experienced much in a short time but none of them ever expected to actually meet Jesus, much less attend a class that he would be teaching.

After the class broke up, Bill's thoughts returned to something that had struck him as odd. Tim had looked at his watch and announced that it was time for lunch. "Why did he look at his watch?" Bill thought to himself, forgetting for the moment the possibility of attending a class actually taught by Jesus. He was starting to get used to Tim's "bombs." If there is no such thing as time, then why does he even have a watch? As he got up and followed the rest of the class, he made a mental note to get to the bottom of this. Tim always seemed to have a reason for everything he did. For the moment, Bill could only wonder what it was.

Chapter Six

Money Class 101
Or How to Get Rich in a Grocery Store

"Good morning, class. This is Money Class 101, and it is a preparatory class for those of you who will soon be returning to Earth Class. If anyone is in here who is not planning on returning to Earth Class soon, you're probably in the wrong room.

My name is Pat Eirdam, and I am your instructor for Money Class. Let's see, this class is scheduled for two hours. That should be more than enough time to make you all millionaires upon your return," Pat said almost smugly. "OK then, if we're all in the right place, let's get started, shall we?

It is probably safe to assume that you all have signed up for this class because you want to have money and enjoy the Good Life when you incarnate next on Earth. It is probably also a pretty good bet that you did not have money in your last, or perhaps even your last several, trips to Earth Class, and that it was a bummer. After all, if you already know how to do it, you wouldn't have any need for this class, Right?

Let me first begin by introducing my credentials and qualifications to teach this class. As you might suspect, I have a master's degree in economics, as well as a PhD in estate planning. I was a Certified Financial Planner licensed in four states and was the United States Secretary of the Treasury twice. I held a chair in economics at one of Earth's most prestigious schools and was the chief adviser to three heads of state…NOT! But hey, lets face it, that's what you would expect of the teacher of this class, isn't it? Of course it is, because you think you need an expert to tell you what to do and what you have done wrong. Well, I've been worth millions in sixteen of my previous incarnations. Once I was even the richest man in Earth Class. And I'm going to tell you how I did it and how easy it is for you to do. In fact, it is so easy you don't even have to do anything"

There was silence in the class. Everyone just sat waiting for the next shoe to drop.

"OK," said Bryan, "that sounds good to me; now what?"

"Now? Now nothing. I don't mean to sound flippant, Bryan, but doing nothing is actually what you do. In your last Earth effort you tried to *do* everything you could think of to get rich and nothing worked; am I right?"

"Yes, looks like you've been reading my mail," Brian offered somewhat reluctantly.

Chapter Six
Money Class 101:
Or How to Get Rich in a Grocery Store

"So this time it will be different, but first you are going to have to unlearn a few things. And the first thing that you are going to have to unlearn is that being rich is not something you do, it is something you are. No amount of *doing* will accomplish anything when it comes to *being* rich.

"You've all heard the poor people say, 'As soon as I get a lot of money, I'm going to do so and so or I'm going to get so and so.' But this is ridiculous. In fact, it is an oxymoron. Why? Because poor people don't have money! They're poor, remember? Only the rich people have money. So the key to having money is to *be* rich, not to *do* anything."

The class sat silently with puzzled looks on most of their faces.

"Look, class, I know this sounds strange to you, but remember, I said we had to begin by unlearning some things. Before we can make any significant progress in this class, you are going to have to understand this one, most basic, concept. Being rich is just that. BEING.

"You have all tried hard, and you've done everything in your power to accumulate money, but no matter how hard you tried, you only seemed to be able to just get by. The operative word here is *seemed*. Some of you even tried to accumulate wealth by, shall we say, less than honorable means. Oh, relax, no one is here to blame or throw stones. But for those of you who, and I don't like to use the word stole, so let's just say, TOOK things, did you ever notice that no matter how much you took, at the end of the day you never seemed to have any more than you did before.

Why do you think that is? I'll tell you why: because even taking is DOING something. Remember, being rich is not about doing anything, it is about being. Besides, the reason you TOOK something in the first place was probably because you believed you were ***"in lack"*** and the thing that you took might help enrich you. I'm pretty sure you never thought you had too much!

"All right, class, lets get into that being/doing thing, shall we? There are two basic choices or approaches that you can take in your upcoming incarnation. The first one is, *you will be what you do*. The second one is, *you will do what you are*. Think about this for a minute. Which one will serve you best? Do you run around doing things so that you will become what you think those things will make you? You could, but what if you get started down a track that doesn't serve you? What then? You *will become* what you practice. So what if the things you practice turn out to be something that will not serve you or you didn't like? Hmmm. I wonder if that has ever happened to any of you in any of your previous incarnations?" Pat said somewhat sarcastically.

"Let's examine it, shall we? Without going into too much detail, the scene would look something like this. You observed a rich person and saw that he did certain things. He had a certain kind of job and went to a particular school. He dressed in a certain way and surrounded himself with luxuries that he wanted. 'That's for me,' you said. I'm going to do everything he does, and then I will be rich too. But somehow it didn't work out, did it? Oh, you bought the same kind of car he had, but you found you couldn't even make the payments. Your other spending only served to put you in debt and your situation worsened. No matter how you tried to copy his moves, nothing worked out the way you hoped it would. Finally, after a lifetime of all this DOING, you ran out of energy and never got what you wanted. It must seem strange to you that no matter how hard you tried, you didn't succeed. After all, hadn't you heard all your life, If at first you don't succeed, try, try again? Well, what happened?"

The class just sat with their eyes glued on the instructor, none speaking, trying hard to understand what they had just heard.

"I'll tell you what happened: You tried to do something to be something, and that is a very difficult proposition that seldom works. If it does work at all, it takes tremendous energy to support and is very delicate, collapsing at the first sign of your inattention. Far easier is to take the second path, and *do what you are***.**

In doing what you are, you are following a much more normal path. You will find your accomplishments attained with far less strain. In other words, in the context of our conversation, *be rich*. Now, of course, many of you are sitting there wondering how you can be rich when you don't have a dime, right?"

The members of the class looked at each other, nodding in agreement.

"I have a feeling you are about to tell us" Brian interjected with a grin in his voice.

"You are perceptive, Brian. I'll say that for you. Yes, class, I am about to tell you. Isn't that why you signed up for this class? To begin with, you must remember that your whole experience on Earth is just to prove to you that what you believe is true. This applies to having money, as well as everything else you will experience there. Remember the expression, If you think you can or you think you can't, *you're right*! If you believe you are poor, then your entire life experience will prove to you that you are right. You will not have enough money in your pockets to buy what you want. Not enough money in your checking account to pay your bills, and certainly no extra money for vacations and other nonessential items. You will always buy the cheapest model car and never travel first class. Now, of course you may not have it this bad. You may believe you have enough to get by and even have a little extra. In that case, your experience will mirror your beliefs. Now I ask you this: why not believe you have all you desire, that you are immersed in a pool of abundance? Wouldn't this be a better belief, since you are just there to prove to yourself that you're right? Wouldn't you rather be right about being rich than right about being poor? I would think so."

"Yes," Sally piped up, "but how can you do that when you know that you are poor and don't have enough money to buy the things you need, much less the extra things you want? I don't see where what you believe can change your situation all that much." She was quite disturbed by the overly simple explanation Pat had offered. After all she had struggled all her life just to put food on the table and provide the minimum for her children.

"Sally, I don't mean to sound patronizing to you, but let's look at what you just said. You said that you knew that you were poor. And I say to you that it was that knowledge that made you poor, not the other way around. If you had known that you were rich, then you would have been rich. Class, this is important. Your life is a mirror of your beliefs and expectations. It becomes what you know. It is what you believe it is. You can't accurately look at your conditions and draw any correct conclusion except that those conclusions are in alignment with your beliefs. Sally, I'm not picking on you here. I know many of the others have the same question in their minds. Believe me, Sally, class, you are not the first group to have this question, and I know it is difficult to accept at first. But it is the only way. As long as you believe that you are either poor or rich your life conditions will mirror those beliefs and you will be what you choose. This class is to help you break the chains of your former beliefs; remember, I said you would have to unlearn some things. If you don't accept this, it won't matter to anyone other than you. This *is* the way it is. It is your choice to accept it or not. Many of you will have other classes that will be addressing this same subject, so I'm not going to go into it in much more detail here, but money, or the lack of it, will color your life's experience to a very large degree."

"I'm sorry, Mr. Eirdam," Sally said. " I didn't mean to sound disrespectful. It's just that I've had several incarnations without money, and I guess I just sort of, as you say, believed it."

"Oh, Sally, I assure you no apology is necessary. Isn't that why you're taking this class? Believe me, I see people in the same situation all the time. The decision you will have to make now is whether your old beliefs still serve you or whether what you have heard here today would be in your best interest when you arrive back on Earth. There is no right or wrong of it. Only what serves you. I have had several people who have taken this class many times and still insist on living a life based in that most destructive of conditions… **lack**. If that is your choice, class, it is just that—your choice! But if you decide that a life of abundance would better serve you, then choose it, know it, be it."

Looking up at the entire class, Pat went on, "So, class, let's get to the second part of this course, shall we? 'How to get rich in a grocery store.' This part will explain, for those of you who are still invested in *do*, how to convert from doing to being. This will be particularly important to you, Sally, and others like you who have lived several lives of lack. Now this is going to sound a little funny to you, but in order to *be*, it is sometimes helpful, temporarily only, to *do*. That doing is an exercise that will be helpful to many of you once you incarnate on Earth and is the essence of the second part of this class. But before we get to that, remember, an incarnation requires a tremendous amount of energy to bring about; therefore, you will want to experience and accomplish a great number of things in a relatively short period of 'time,' as you will refer to it on Earth. Some of you will simply be born into a rich family to assure your indoctrination in the abundance theory. Most, however, will choose an experience rich in other aspects; therefore, you will be born into families that will have a focus other than money. But you did take this class and you will have an interest in abundance. However, since your parents won't have money, you will have to 'remember' this class and teach yourself how to live in abundance. And you will remember it in your subconscious. Your upbringing may have tried to make you believe that you simply must accept your lot in life. For many of you, the mere fact that you have elected this class means that you will reject that theory and shift your focus to abundance. In order to do that, you will have to teach yourself how to do something. And that something is coming right now. So let's go grocery shopping, shall we?

"Now remember, class, the truth of the matter is, as I have said over and over, you don't really have to do anything. I can't say this enough. But if you find yourself trapped in the belief that you are a victim and that things can happen to you that you are powerless to control, which of course is not ever true, then this little grocery-shopping exercise can be very helpful. Now, as an exercise, by its very nature you will be doing something, but this doing is only a temporary situation until you see that you can be in a condition without having to do anything to get there.

"OK, lets get started, shall we?" Pat continued. "I'll set the stage for you. You are thirty-five years old, married, with three children. You had a somewhat middle-class upbringing, but your parents were always struggling to provide a living for themselves, you, and your brother. They were believers in saving for a rainy day, which seemed to occur all too often. You were always cautioned to be conservative and demand value for your dollar. As you grew older you honed that concept to an even sharper point. You shopped at discount stores, even if it meant traveling out of your way. You were charitable, especially at Christmas time, but you always had the feeling that whatever you gave away was just that much less for you, and, as in almost everything else, you were conservative. You didn't often give money to bums, as you thought they would not use the money wisely. Of course, you wouldn't do that today, knowing what you know now about the power of giving.

"Anyway, as the years passed and your family grew older, the demands on your money grew more acute. As a result, you became even more frugal. When you went to the grocery store, you saw no sense in buying the deluxe and more expensive brand-name products. You felt a need to stretch your budget as far as possible. Without actually realizing it, you slipped into the belief and under the influence of **Lack**. This is where you are going to have to remember this little exercise. Don't worry how you will remember it. If you have the desire and you choose to experience your desire, it will come to you. Maybe through something someone said to you one day, maybe through an inspiration you had. Heck, you may even read it in a book, but no matter, it will all only be coming from you and the time you spent right here in Money Class. Seriously, though, it will bleed through. And when it does, it will just seem right, and you'll recognize it as the answer you're looking for.

"OK, the stage is set; you're in front of your local grocery story, and you're about ready to go in. Your life is about ready to change forever.

Chapter Six
Money Class 101:
Or How to Get Rich in a Grocery Store

"The first thing I want you to do is stop in the parking lot and just look at the store for about fifteen seconds. After all, this store is just about to change your life. I want you to realize that there is nothing in this store that you cannot buy if you really want to. Individual items in a grocery store are within the means of just about everybody. Say to yourself, 'I can buy anything in that store that I want.' Oh, sure, you can't buy everything, but you can buy any *one* thing you want. I mean we have to start somewhere.

"All right, you're in the store now, so let's go up to, say, the paper-towel aisle. Look at all the paper towels. You see your brand; it's seventy-five cents. And you see the most expensive brand, and it is $1.50. 'No!' you cry out, 'surely you're not going to make me pay double for a roll of paper towels, are you? Paper towels are made for throwing away. Why would anybody pay double for something that is just going to be thrown away? Oh, all right, this is your exercise. I guess I'll go along with it,' but it is painful to you in any case.

"Now you get those towels home and start to use them. 'Wow,' you say, 'these babies are really soft, and so absorbent, too.' For a moment you are even tempted to lay them out to dry so that you can reuse them, but no, this is, after all, an exercise in abundance, and acting frugally at this point would only serve to lessen the experience if not negate it altogether. So for the rest of the week you continue to enjoy the quality and performance of these superior paper towels. They really are nice, much better than your old cheap brand.

"Finally, you are getting low, and it's time to go to the grocery store again anyway. Once again, you come face-to-face with the paper towel aisle. Once again, you agonize over the price of the premium brand, but you are determined to see this experiment through and choose it once more. Once more, you experience the benefits of the premium brand.

"You will notice, class, that each time you return to the store for paper towels, the decision to select the premium brand becomes easier and easier. Soon, you don't even look at the price; you just automatically select *your* brand. You are the king/queen of the paper towel aisle.

"You haven't experienced any undue hardship for the purchase. The money was somehow there for it. You're no worse off. In fact, you're better off. You now are a premium towel user. No millionaire could buy a better brand. When it comes to paper towels, *you're rich!*

"This is your first step towards not accepting littleness. You are, in pure fact, too big to be contained by small ideas. This you must learn in everything you are and, yes, do!

"Any questions so far?" Pat looked at mostly blank stares as the students struggled to make sense of all they had just heard.

"All right then, lets continue our shopping spree, shall we? Let's go down the, ah…bean aisle, OK? Now here we are confronted with the same scenario—cheap beans or premium beans. We already know which decision we are going to make, and it is easier this time. After all, we weren't plunged into poverty with our paper towel extravaganza. Soon, just like with the paper towels, we are the king or queen of the bean aisle.

"Now, you're pretty sharp folks here, and you can see where this is going. You just keep applying the same concept to all the aisles in the store until you are the king or queen of the grocery store."

Raising her hand again, Sally waited for Pat to acknowledge her. Pat seemed as if he was waiting for something before inviting Sally's question. Finally he spoke.

"Yes, Sally," he said.

"I can see how the technique you just explained would work for any single item or even several items, but if you continued to do that with every single item, by the time you got your basket filled it would be too expensive to afford."

"What do you think about that class?" Pat said, making eye contact with each one. This was the reason he hesitated before calling on Sally. He was hoping she wouldn't be the only one to express this objection or at least think of it.

"I'm not sure," Bill said after a while.

"Go on, Bill, why not?"

"Well, if the money was there to buy the first item, or as Sally said, even the first few items, then the same principle should apply to the whole basket, I think," Bill added tentatively.

"Very good, Bill, and you would be correct," Pat said.

"Can you explain why, Wayne?" Pat said, looking at Wayne.

"Ah…well, I guess it's because by the time you get through the exercise for each item, you would have seen that you did have the money to afford each one, and I guess you would just somehow have the money to afford the whole basket. I don't know. I guess I'm not sure."

"No, no," Pat said. "You've got it. That's exactly right. You would have the money."

"But where would it come from?" Sally asked. "I guess I just can't understand how the money would, well, just sort of magically appear."

"Oh, it wouldn't be magic, Sally," Pat said. "The money wouldn't have any other choice."

"I don't understand. I guess I'm just too dense or something...," Sally said, her voice tapering off.

"Not at all, Sally. Look, what are we trying to do here? We're trying to move into a *state of being* and our objective is to be rich. Therefore, once you know you are rich, you will be rich. Every rich person knows they're rich, just as every poor person knows they're poor. The difference here is that the rich person never questions that he has the money to buy the whole basketful, and we're trying to get to that same state of knowing, one item at a time. Each time you add an item to your basket, you observe that you somehow managed to get out of the grocery with it. You did have the money. It did happen. Remember, everyone, the physical conditions that you find yourself in are in direct correlation and in response to your beliefs about that situation. In other words, what you know, much more than what you believe, is what will be true for you. Now, if you know that you cannot afford the items in your basket, then that exact situation will manifest itself in your physical reality. You will not be able to afford them. But the object of our exercise is to condition you, one item at a time, to the fact that you can, in fact, afford the whole basket.

"Let me explain it another way. In a very real sense, it is not *your* responsibility to get the money. After all, that would be *doing* something. Your job is to accept that your very request is a command, and all the forces in the universe conspire to obey it. At first, you may not be aware of this. In fact, you probably won't be, but soon, little by little, after observing that your commands are obeyed, you will come to know that there is no other way.

"One of the main purposes of a physical existence is to provide the environment where you can observe the effects of your power. Someday, this concept will not be difficult for you to understand, but rather will be self-evident, and you'll hardly remember a time when it wasn't so. Therefore, class, do not be amazed when your commands are obeyed; rather learn to expect it. However, I caution you, if you believe that your commands will not be obeyed, then that in itself is a command and *it* will be obeyed. Where the money comes from, at its original source, is actually a bit beyond this discussion. I only ask that you use your power to command your every wish forward. You will do this, however, by choosing it, not by wishing for it. You *must* choose it. If you only want it, then that will be what you'll receive, *the want of it*. Remember, class, everything already exists in a less-than-physical form; it is your job to excite it to a physical form, and that is done by a command. A command backed by the knowledge that there is no other possible outcome. Everything created by man has first been a thought, then a command, then a physical manifestation.

Whether you understand it or not, everything is already so. It already exists; it only awaits your command to spring from your thought to the physical object that you desire. The time that it takes to accomplish this waits only on your degree of understanding of the knowledge that it will occur."

Sally looked somewhat dejected as she said, "I understand what you are saying now; it's just that I'm not sure I will ever be able to make it work."

"Don't worry about it, Sally, you've got a very long time—eternity I'd say—to work it out. I know you'll make significant progress on this on your next trip. You'll see," Pat said, and there was a tone of genuine affection in his voice as he smiled at Sally.

"You all will be having upcoming classes that are designed to show you that you're bigger than your dreams, not the other way around.

"Well, that's about it for today, unless there are any questions. I want to thank you all for coming." As the rest of the class got up and left, Wayne remained seated in the back of the class.

Pat, gathering his papers, looked up to see Wayne, still sitting, with a concerned look on his face. Pat stood up to his full height, took in a deep breath and folded his arms across his chest, then leaned back against his desk. There was a moment of silence between the two men while each looked at the other. Wayne was the first to speak.

"Excuse me, Mr. Eirdam, I think I've got most of this, but I have one question," Wayne said, speaking very deliberately, as though choosing his words carefully.

Chapter Six
Money Class 101:
Or How to Get Rich in a Grocery Store

"Earlier this morning, when you were setting the stage for our shopping adventure, you said something that didn't quite fit, at least for me. You said in your example that we didn't often give money to bums, because we didn't like what they did with it. I've been wondering why you said that. You don't say things just to be saying something. What was your point? Shouldn't we be giving money to bums? I understand that we should be charitable, but aside from that, if we give something away, how can we possibly have it? I mean, this is Money Class, and I should think it would focus on getting money, not giving it away."

"Whew, that's quite a question, Wayne, or should I say questions? You are correct; this is Money Class, so I guess I owe you an explanation, though I didn't really want to get into this right now. You see, Wayne…," Pat said as he hesitated, knowing his next few words were going to open up a whole new subject, "the real reason that you will not be diminished by giving something away is because there is only One Person in Earth Class, and that Person is God. I know how this sounds. I know you *see* many other people, and this concept has to seem very confusing, but those other people that you think you see are just aspects of that One Person. So you see if there is only One Person and you are an aspect of that One Person, then giving to any other aspect of that One Person is the same as giving to yourself. When you give to yourself, you cannot possibly be diminished. No candle is ever diminished by lighting another.

Wayne, this is really a very large subject, and it will be referred to in many of your upcoming classes, but as it pertains to Money Class, I'll say this. Have not the greatest Avatars who ever lived— Buddha, Jesus, Muhammad, Spalding, to name a few—all had the same message? That it is preferable to give rather than to receive. This is not a recommendation for the conduct of a *good person*, It's the essence of how best to get what you want. Think about it for a moment. If there truly is only One Person in Earth Class, who else could that Person be giving to except Himself. Therefore, since you are an aspect of that One Person, whatever you give to another, you can only be giving to yourself. I would be remiss, however, if I did not tell you that to withhold from another has the exact same effect in the opposite direction, and to do so could leave you in a pretty sorry state as far as your financial health is concerned. Therefore, I would heartily recommend that you give to yourself via the avenue of giving to, what appears to be, another person."

"Wow," Wayne said. "That's heavy. Do you mean that when I want something, I should give it to another person, thereby getting it myself?"

"Well, not exactly. You see, if your purpose in giving to another is to get it for yourself, then you have not truly given. You have a string, so to speak, attached to what you gave. As the receiver, you will perceive this gift as not quite attainable. You won't see the string of course, but you will feel a resistance when you try to possess the object of your desire. It will be difficult to attain. And why? Because it has never been truly given, and you cannot fully possess something that is still possessed by another. But when you give freely, with no thought of return or benefit, you cannot help but receive, because, in a very real sense, you are only giving to yourself. Again, though, you must give to yourself completely, without reservation, so that you may receive completely without limitation."

"That's a lot to think about, Mr. Eirdam," Wayne said, and there was a look of wonder on his face.

"I know, Wayne, but you will understand it better after some of your upcoming classes. Heck, soon you'll wonder how you could have thought it was ever any other way."

Chapter Seven

Knowing—The Secret to Everything
Or Learning How to Swim

"Good Morning, class, it's nice to see all your eager faces this morning. Today's subject is learning how to swim." The class looked around at each other, not sure they had heard right or thinking that perhaps they were in the wrong class.

"Those of you who have had physical lives may wonder what you are doing here, especially those of you who already know how to swim. I assure you there is no mistake. This is a required class and a very important one at that. In this class, you are going to learn how to shortcut the process that you have referred to as learning."

Still, the entire class sat with looks of quiet puzzlement on their faces. Tim's explanation so far had only served to add to their confusion.

"Those of you who are using this trip to Earth as your first physical experience could greatly benefit from what I'm going to say here today. But please notice I said *could benefit*. There have been many people before you who have attended this class and still have not succeeded in perfecting this process. As you are all now starting to learn, your entire physical existence is formed by your mental expectations. This process, and a process is what it will be for most of you, will help you see just how your mental projections create your physical surroundings and conditions. Physical process generally proceeds fairly slowly for those of you who are new to the physical experience. And while that may seem undesirable, as you will soon see, that is not necessarily so. In fact, at times it will offer you protection exactly because there is a delay between what you think and what you manifest in the physical. I know that seems strange to those of you who have not had a physical experience. Your classmates who have had a physical experience will attest to what I am saying. For them, this class is actually optional. Most of them have selected it to get some additional help as to just how the whole process works so they will be able to use it more beneficially.

"So, as we get started, remember the most important thing, and I mean *the most* important thing, for you to keep in mind is this simple fact: *Your body and your entire physical existence will be nothing more than your thoughts manifested in a form that you will be able to see*. In fact, that's what a physical existence is. Now, having said that, let me explain how it works.

Chapter Seven
Knowing—The Secret to Everything
Or Learning How to Swim

"Shortly after your arrival on Earth and followed by your brief but very important childhood, you will start becoming aware that your actions are actually influencing the events of your physical existence. You will notice that certain thoughts produce certain actions. You will start to organize and direct your thoughts as you see them manifest into various actions. You'll like some of the outcomes and dislike others. Soon you will make a direct effort to control the outcome. This you will call planning, and you will start making plans for much of what you do. You will notice that the objects of your desires will manifest themselves much faster and with much more ease when you carefully execute and shepherd your plan.

"Some of you, especially those who have been through this before, will become quite proficient at converting your thoughts into physical experience. For you, execution of your plan will seem simple, and life for you will seem easy and natural. Your friends may even call you lucky.

"Others of you, however, will struggle through your entire physical experience without ever grasping the basic principles. For you, it will seem that life is subject to and full of uncertainties, and *things* will seem to happen to you that you did not direct.

"Of course, sitting here in the comfort of this classroom, that notion must seem absurd to you, but again, ask your classmates who have been physical before how really easily it can happen.

"As I said at the beginning, the purpose of this class is to show you how to shortcut the process of learning and to explain the exact mechanisms by which the experiences and conditions of your life will be created. So let me get back to the subtitle of this class: **Learning How to Swim**. Of course, the process applies to learning how to do anything, but the example of swimming serves quite well. Notice I said how to *do* anything. I would have been more accurate to say learning how to *be* anything. Being is really a much more accurate statement when it comes to learning than doing is."

Several of the students shook their heads in obvious confusion.

"OK, class, I know it sounds a little confusing, so let me start putting it together for you right now. Let's say that you are out at a lake on a nice summer day with some of your friends. Everyone is having a good time, and many people are swimming. Eager to join them, you walk out to the end of the dock and jump in. But wait! Do you know how to swim? Now is not really a good time to get the answer to that question, especially if the answer is no. In fact, you wouldn't jump in until *after* you had determined whether or not you knew how to swim. But how do you determine that?" Tim asked no one in particular.

"Well, I guess you could ask the question as you jumped off the dock. The outcome would be fine as long as the answer is yes. But how can you be assured the answer is yes?" Tim said, and paused, looking over the class.

"The answer is simple. By *knowing* that you can swim." Then he paused, gauging the reaction to his last comment.

"Many of you who have not had a physical existence yet are thinking, well, that's simple, and yes, in fact, it is. Unfortunately, some of you will have forgotten this class, and by this time your well-intentioned parents will have pretty much convinced you that you will drown if you go into the water without first having gone through a ritual. That ritual, in our case, is called swimming lessons. And it involves a process. I could just as easily have used the term learning. Process and learning are interchangeable words. They are the entire reason that you will use the highly illusionary concept of time."

Several of the students looked up intently with questioning frowns on their faces.

"No, no, we're not going there right now," Tim said. "The concept of time is a difficult one in itself, and while there is no such actual thing as time, it would be too distracting to digress into that subject right now. However, I can assure you that you will get an exhaustive explanation of time, err…rather the lack of it, before you leave for Earth Class. OK, back to our swimming lessons. While the process itself is of no importance, your belief in it is. You believe that you must go through it in order to know it. This is where you use time. As you have heard many times before, time doesn't really exist except to serve process/learning. You see, class, you must have an explanation for your new talent, swimming. It will be too difficult for most of you to simply accept knowledge. Most of you will come to believe that you must earn it through process. Of course, this belief is not true and will limit your experience, but if you make it true, you will make it real, and, therefore—the need for swimming lessons. Now, you set off to learn how to swim, but lets put in a little diversion here, shall we? You are a cautious person, so before you actually start your lessons, you decide to get a complete physical. Ah…that's an examination by a doctor, not to be confused with the act of being physical…. Oh well, never mind. So you go to the doctor and say, 'Doctor, I'm going to take swimming lessons, and I want a good examination to see that I am up to it first.'

" 'All right,' the doctor says, and starts the exam.

In a few minutes, he pronounces you fit and ready for your lessons.

" 'But wait, doctor. I was hoping for a more thorough exam. I'd like a full-body X-ray and a brain scan and blood tests and stuff like that,' You insist.

" 'Well, that's a little much under the circumstances,' the doctor explains. 'Your insurance company most likely will never pay for such tests.'

" 'That's all right!' you exclaim. 'I want them anyway. I'll pay for them myself.'

" 'All right,' the doctor says, and sends you off for the tests.

"In a few days, the results are back, and, sure enough, just as your doctor had foretold, you are fit as a fiddle. So off you go to your first swimming lesson.

"The first day is a little discouraging. You can't seem to accomplish even the simplest of techniques, but no matter, you are determined, even though the second day doesn't fare much better. On the third day, you manage to accomplish your first few strokes, even if they are pretty much by accident, and you are pretty pleased with yourself. The fourth and fifth days produce ever-increasing success, and, for the first time, you think you may actually learn to do this thing after all. On the sixth day, the *breakthrough* occurs, and you have mastered the basic strokes. You are no longer afraid to be in water over your head, and you can make it all the way across the pool without too much effort. Finally, at the end of the week your instructor pronounces you a swimmer.

"That night at home, you start to think about what has happened to you that caused you to swim. You decide to go back to the doctor and see if he can explain why you are now a swimmer, when only last week you were not.

"The next day you burst into the doctor's office announcing, 'I can swim, I can swim!'

"'Wonderful!' the doctor says, and there is a tone of sarcasm in his voice.

"'What can I do for you today?'

"'Well, doctor, I want to find out what has happened to me.' And there is a sense of urgency in your voice. 'Could you do another x-ray of me for comparison? I need to know what has changed. What has happened.'

"Well, of course by now the doctor thinks you're a little strange anyway and informs you that he has no answer for you, but is convinced that any additional tests will fail to show any difference from the previous ones. So now, class, does anyone know what happened?" Tim asked."

Bill raises his hand.

"Yes, Bill," Tim said, acknowledging Bill's raised hand

"Through the swimming lessons, he learned how to swim," Bill said. "He just kept practicing and practicing until he got it."

"OK, fair enough, Bill, but why must our swimmer practice over and over? Why didn't he just do it right the first time after his instructor demonstrated it to him?" Tim inquired.

"Because, ah, because, well...," Bill stammered. "Because it takes practice to work it all out. You can't just do it right the first time you try something."

"You can't?" Tim questioned. "Why do you think that?"

"Well, because, because that's the way it is. I don't know."

" I see," Tim said, folding his arms across his chest and walking back to the head of the class. "So he had to practice over and over because that is the way it is, huh?

"Well, it's a good thing You're taking this class today, because you are going to need it if you want to shorten the 'time,' as you'll come to call it, that it takes to learn.

"All right, class, here's what happened. Through a process—that process that we call learning—you came to know that you knew how to swim. If our good doctor had taken additional tests, you would have seen that you didn't grow a swim bladder, your neural net didn't completely reorganize itself, and you didn't grow webbing between your fingers and toes. The only thing that happened to you was mental. You moved into a position of knowing. You knew that you knew how to swim. You only used the process and the 'time' it took as a tool. In reality, you didn't need either. All you needed, all you'll ever need is the understanding that you know. And that is the only difference between the person who jumps off the dock and one who doesn't. The person who jumps off the dock knows he knows how to swim and doesn't even think anything of it. No one jumps off the dock and then, on the way down, says, 'Gee, I wonder if I know how to swim.' If you have to ask, you don't know how.

"So, for a while you need the process to validate your new belief. You need to know that you did something. In reality, you need do nothing, just move past belief to knowing. This process is often called *self-consistency*. It gives you a path whereby you can explain to yourself how it came to be that you could now do something that you once could not do. You remain *self-consistent*.

"This is one of the biggest uses you have for that stuff you call time. Even though things will change in your Earth Class experience, you will understand how you fit in and what they mean to you, because of the passing of time. Your body will get older, but it will still be you. You may find yourself in a different situation, such as a different job or even a different life partner than the one you started off with, but you will understand how it came to be, because of your use of that illusionary stuff you call time. You will at all 'times' remain *self-consistent*. It doesn't matter what you call it or how you try to explain it to yourself; the same mechanism is at work, and it's just there to give you the self-confidence you need to move from a state of wanting or desiring into a state of being and knowing. And that, dear class is what allows you to jump off the dock without fear.

"Which leads me to the introduction of your next class—FEAR, the one and, by far, most potent obstacle to your development, both in and out of the physical. But that's for tomorrow, so…"

Chapter 8

Fear and Guilt
Or Why You Drag That Anchor Around

"Good morning, class. My name is Philip. Actually, that's not my real name, but you couldn't pronounce my real name if I told you, so Philip will do just fine."

The new instructor standing before them was fairly short, rather rotund and dark complexioned, and reminded Wayne of an East Indian in both his appearance and speech.

"This morning, we are going to discuss in detail a subject that, more than any other, has kept you from realizing your true and full potential: Fear! Anyone care to venture a guess why fear is so destructive to an Earth Class experience?" Looking around the class, Philip's eyes settled on Wayne.

"Wayne, how about you? What do you think about it?"

"Well," Wayne said, surprised to be the first one to be called on, "I guess it's because if you're afraid, you will do less and, hence, experience less. At least that's what I'd say."

"And why would you say that, Wayne?" Philip asked with a tone one uses when one already knows the answer.

Drawing a deep breath and looking somewhat reluctant, Wayne started to speak. "Because…that's exactly what happened to me. I see that now," he said, talking more to himself than answering Philip's question.

"After listening to the other instructors and hearing how things could have been different, I wish I hadn't been so fearful. Things are different now, though." Wayne's voice had gathered some volume, and he sat upright in his chair. "Things *are* different now!" He repeated.

"If I'd known only half of what I've learned here in the past few days, I could have been twice the man, no, ten times the man I was. But…" His voice trailed off, his shoulders drooped, and he slouched in his chair. "It's too late. That's all in the past. There is nothing I can do about it now."

"Wayne, I think you are living under a misconception," Philip said.

"Before we get too deep into the subject matter of today's discussion, I believe we are going to need a bit of a review here. How can there be such a thing as 'too late' when there is no such thing as 'time' in the first place?" Philip mused as he looked over the class waiting for the rebuttal that he knew was sure to come.

Chapter 8
Fear and Guilt
Or Why You Drag That Anchor Around

Wayne looked around the class, not wanting to be the only one to talk. Surely he wasn't the only one who was having trouble with this whole time bit. Alas, no one else was going to speak. Philip was already looking directly at Wayne as he gave in and said, "I'm sorry, but I'm still having trouble with the concept that there is no time. I mean, it just seems so real and everyone experiences it."

"Yes, Wayne, I know it appears that way, but what have you heard here over and over? That physical reality is an illusion that *you* make real. Let me ask you a question, Wayne. How old is the Earth?"

"Ah…I guess I don't actually know?" Wayne said, looking around the class as if for help.

"We'll ask our scientist, shall we?" Philip said, looking deliberately at Mary Beth.

"About 4.5 billion years," Mary Beth quipped with hardly a thought.

"And the universe?" Philip asked.

"About 14 billion years, maybe a little less," Mary Beth said matter-of-factly.

"Would you agree with that, Wayne?" Philip asked.

"Ah, sure," Wayne answered with a 'why-not?' expression on his face. "I'm sure Mary Beth knows this stuff."

"OK…all right," Philip said, putting his hand up to his chin as he strolled across the front of the classroom.

"Do you agree with that, Alan?" Philip said suddenly, turning on his heel and addressing the young man sitting next to Wayne.

"I couldn't say, sir," Alan answered. "As you know, I have never been to Earth Class, and I have no idea how long a year is. My only physical experience was on a planet a long way from Earth Class, and we didn't use that particular measurement."

"But you did note the passing of time, did you not?"

"Yes."

"What can you tell Wayne about how you did it?"

"I'm not sure. There doesn't seem to be any common measurement between our two experiences," Alan ventured.

"I see. And herein lies our dilemma *and* our answer," Philip announced rather smugly.

This latest revelation did little to answer the question as far as Wayne was concerned.

"You see, class, time does not 'flow.' I know it appears to start in the future, flow through the present and out the back door to the past, but that is an illusion. For one thing, it all depends on how fast you are going."

Again, this explanation did little to unravel the mystery, and several people shifted uneasily in their chairs.

"You see, class, the faster you travel while you are in a physical condition, the slower time appears to pass. A noted personality named Einstein, in a recently completed Earth Class, was among the first to notice that the closer you get to the speed of light, the slower time seems to pass. If you were to ask a person who was traveling at near the speed of light how old the universe was, he would report that only a very short time had passed since the creation of the universe, perhaps only minutes. If you were to ask a person traveling faster than the speed of light, he would report that he has no knowledge of such an event, and, hence, the universe hasn't started yet. The only reason that you say the universe is 14 billion years old is that you have noticed that the sun in Earth Class rises every twenty-four hours, an arbitrary time increment in itself, and you have decided that when you accumulate 365 of these risings you will call it a year. Further, you have assigned 14 billion of these years to denote the 'beginning' of time. But as you heard, Alan does not know what you are talking about. The physical experience he just completed occurred at a much faster speed, and if he were able to extrapolate his time reference to yours, Wayne, he would say that the universe was something on the order of one thousand years old. So you see, there is no actual definite increment of time. Different people traveling at different speeds would report widely varying concepts of time."

"OK, I can see that, but they still would report the passing of time, no matter how they experienced its interval, wouldn't they?" Wayne said, his eyes wide open and staring at Philip for the answer.

"Yes, they would, Wayne, they definitely would."

"Well then, doesn't that mean that time exists, even if it is different for different people?" Wayne questioned.

"NO!" Philip said abruptly, pausing to let that soak in. "It doesn't."

Wayne flopped back in his chair with a defeated look on his face. He shook his head. "I just don't get it," he said.

"I said time does not exist; I never said it doesn't appear to. Of course time appears to exist or else you couldn't make sense of your experiences. One day you would appear to be eighteen, doing all the things that you would do at that age, and the next you could appear to be sixty-five, doing all the things you would do at that age. The next day you could be in elementary school, followed by the next day finding you in a foreign country on a vacation. Without the appearance of time, you could not make sense of your physical experience. All of the experiences I just mentioned would be you, but you would not be able to understand how you got into those different situations, particularly if it appeared that you had traveled backward in time, say going from an old person one day to a young person the next. So to make sense of your experience, you invented the illusion of time. Now everything can make sense. Actually, events in your life could more accurately be described as little event boxes, looking more like the individual frames on a movie film. Just like time, all of the frames exist at once; they are just chosen one at a time. The little boxes are not actually moving, you just *appear* in different boxes, creating the illusion of time, to say nothing of the illusion of movement. Just like the way a motion picture film works, you are in one box, and then you are in another. Nothing actually moves, and no time passes, you just focus your attention on different frames. To make sense of the experience, you use the illusion of time to separate the events and make them appear chronological. It would be more difficult for most of you, at this stage in your development, to make sense of your Earth experience if you jumped forward then backward in time. Some would lose their sense of *Self-Consistency*. Now, I said it would be difficult, I didn't say it would be impossible, and many people in Earth Class, particularly *teachers*, do just that all the time.

"Speaking of going backward in time, let's go back to our point; so much for the quantum physics lesson.

"You said, Wayne, that it was too late, referring to the fact that you believed that you could not go back and undo some of the events in your just-completed Earth Class experience. Events that you implied were more limited than you would have liked because of your fears; is that correct?"

"Yes," Wayne answered, and there was piqued interest in his voice. "If only there was a way." He waited eagerly for Philip to continue.

"What if I told you that you *can* go back and do things differently? What would you say to that?"

"I'd say show me how!" Wayne blurted out, and there was excited anticipation in his voice.

"All right, but I don't really have to show you anything. You already know how; you've done it many times already."

"I don't understand," Wayne said. He desperately wanted to hear how to do this, but it sounded like this could to be another long, drawn-out explanation that all the instructors were so fond of.

"Would you please tell me how?" Wayne pleaded, hoping to circumvent any lengthy explanation.

Philip only chuckled. He could sense Wayne's impatience.

Philip started to speak. He knew there were many ears waiting for this explanation other than Wayne's. This was a subject that never failed to arouse interest.

"Class, when you do something, or you don't do something that you wish you had done, there is always a chance to redo it. After all, isn't everything you do merely a choice? Isn't it just one of the possibilities of many that you could have chosen? You choose the one that you believe will serve best for reasons that seem valid at the time. Later, you may wish you had chosen differently. As long as you believe that your choice was cast in concrete and it is too late to choose differently, you will not even try, believing there is nothing that can be done. But remember, before you made the choice that you did, there were other options. I submit to you now that those options still exist. Let me give you an example. Say a friend calls you up and asks you to go with him to the movies. For the sake of our example, you have three choices: 1. You can say, 'yes, I will go with you to the movie.' 2. You can say, 'no, I will not go to the movie with you.' 3. You can say, 'I will not go to the movie with you, but I will go,' lets say, 'roller-skating.' Now, until you make a choice, would it be fair to say that each option is legitimate and each could be chosen?"

The class nodded in agreement, looking at one another.

"All right then, you agree that each choice is just waiting to spring into reality? It only awaits your decision? Each can be done, agreed?"

Again, the class nodded its agreement.

Chapter 8
Fear and Guilt
Or Why You Drag That Anchor Around

"OK, then, let's choose number 1. Now, at the moment of choice, number 1 springs from being only a potential choice to an actual event. But wait. What happens to number 2 and number 3? If they were valid before you made your choice, what do you think happens to them? Do they just 'puff' vanish into thin air?" Philip hesitated before he continued, eyeing each student individually.

"If they were 'real,' what now makes them less than real? Your choice you say! All right then, if it's only your choice that made number 1 real, then it should be a simple matter of choosing again to make one of the others real. Make sense to you?" Philip again eyed each class member.

Wayne took a deep breath as if he was going to speak, but only sighed, not knowing exactly what to say.

"Now, I know what you are thinking," Philip said in his best Clint Eastwood impersonation. 'How do I redo choice number 1 and choose choice number 2 or choice number 3?' Well, the choosing is the easy part; you just do it. The hard part is realizing that you have re-chosen. In fact, you *cannot ever realize it*."

The class just sat silently. No one moved.

"After you re-choose, you are not aware that you have re-chosen. You simply experience the new choice and are not aware that you had previously chosen something different. The experience of roller-skating is simply 'different' than the experience of going to the movie. When you go to a movie, you do not sit there and consciously think that this is different than roller-skating. You simply accept the fact that you chose to go to the movie and enjoy that experience. Now, when you re-choose, you go down a different 'timeline,' and for you, the first choice no longer exists. It is sort of erased, for lack of a better term. Your life experience seems to just flow from one event to the other without you ever being aware of the fact that you have re-chosen many events many times, but you do, and you have. Therefore, Wayne, the answer to your question is simply for you to choose what you want, as it always is. You will not be consciously aware that this may not be your first choice and that you are re-choosing. The event will appear to you as if you were experiencing it for the first time. Just like the choice of whether or not to accompany your friend to the movie, your choices will be infinite, always. You are always free to choose again. Fear is like everything else; it is a choice. It is something you must practice to get good at. It is created and develops itself when you make the same fearful choice over and over again until it becomes a habit. Eventually, you don't have to choose at all. The mere thought of the fearful activity is enough to induce a state of anxiety. Remember, class, 97 percent of everything that you will fear and worry about in Earth Class will never come to pass. Fear is just a tool. It is there to push against and to overcome. Overcoming it, like everything else, is a choice. Don't forget, you had to practice to learn how to be afraid. To overcome fear requires the same thing—practice.

"If you choose not to overcome fear, your experiences will be limited by it, as limitation is its only purpose."

David's fidgeting in the back of the class did not go unnoticed by Philip.

Chapter 8
Fear and Guilt
Or Why You Drag That Anchor Around

"David," Philip said, looking over the top of the reading glasses that always sat on the tip of his nose. "Why do I feel that you're not in complete agreement with the discussion here this morning?" David had already been through several classes, but still he was just too reluctant to take a stab at what would be his first physical reality.

"Oh, excuse me, sir," David said. "It's not that I don't agree with what you are saying, it's just…well, it's just too scary. I don't think I could do it. I mean, why would I want to do it? I've listened to all the instructors, and I find all that uncertainty that can occur in Earth Class a little daunting, if not downright scary."

"Really," said Philip, sitting down and leaning back in his chair. "Why is that?"

"Well, for one thing, how can you control anything in Earth Class? I mean, you can't. Unless I've missed something here, as I understand it, you could be in the crosswalk, the light could be green and you could *still* be run over. That shouldn't happen. That's against the rules, isn't it? I've been hearing that things happen that shouldn't happen all the time. Wings that are never supposed to come off an airplane do sometimes. Bridges that are designed to last for many years sometimes fall down long before their time. How does anyone know what to do? How can anyone assure his safety with all that uncertainty? And that's not all. How can I be assured that I will even live to reach adulthood. Some parents actually kill their offspring." David was getting really wound up now. There was genuine fear in his voice, and his words were coming faster and faster. "I could catch a disease from someone and not even know it until it was too late. I could become disabled and, and…even feel pain. And…and…"

"David, David!" Philip interrupted with a somewhat raised voice. "It's all right. You don't have to go. No one is going to force you to do anything you don't want to do."

"I'm not going! I'm not going! I've made up my mind!"

"All right, David, that's settled." Philip paused to let everyone catch their breath. "Are you OK, David?" Philip asked, and there was a calming tone in his voice.

David nodded. "Yes, yes, I'm OK," he said

"Good. If you wouldn't mind, though, I have a question I'd like to ask you."

David looked up. Both men's eyes met, and there was a short silence.

"Why did you sign up for Earth Class if you didn't want to experience physical-ness?"

"Because I didn't know until I got here how scary and uncertain things were going to be. I did want to try a physical experience, but now I'm afraid of it. It just doesn't sound like it would be very enjoyable," David said, relieved at not having to go.

Philip only smiled.

"Gee, David," Bill interjected, sitting up in his chair. "It's too bad you feel that way. I mean, sure, all those things that you just mentioned could happen, and in fact they do, sometimes, but believe me, it's not like it happens all the time."

"I don't care, Bill," David said in what was now a calm voice. "I just don't see any reason to take that kind of a chance if you don't have to, and I *don't* have to!"

Chapter 8
Fear and Guilt
Or Why You Drag That Anchor Around

"No, David, you don't, Bill said. But believe me, you will be missing one of the most incredible experiences you could ever imagine. You cannot comprehend the richness of a physical experience. It's just a shame that you are letting your fear rob you of it."

Philip stood up from his desk and walked over to the window, folding his arms across his chest as he went. Finally, he spoke. "David, Bill is right, you know. It is a fantastic experience. But then you know that yourself, deep down, don't you?" David sat uneasily, not saying a word. "It is fear that is holding you back, and that isn't news to you. After all, that is its only purpose. You heard Wayne say how much better his own experience could have been if it had not been so dominated by fear. You've just taken it one step further, that's all. You're too afraid to even want to try it. And you know something, David? I'm on your side on this one. I don't want you to try it, either!" Most of the class by now realized that there was going to be more to this than Philip was letting on. "Do you know why, David?"

"No, I can't say that I do," David responded with a tone that indicated he wasn't sure he wanted to hear what was coming next.

"Because, if you go there now with the fearful feelings you have chosen, your experience will be one of fear. After all, you will only experience what you believe is true. That alone will hold you back from experiencing all that it could be. Am I right, Wayne?"

Wayne sat up taller and looked directly at Philip.

"Is that what happened to me? Wayne asked, and there was a tone of deep questioning in his voice. "Was I that fearful when I left last time? Is that why my experience was so…so limited?"

"Yes, Wayne, that's correct, and I have to accept responsibility for that. You were in my class. I had just started teaching then. It was my first class, and I was eager to help you. I was sure that you would see what a wonderful experience Earth Class was. I thought you would be all right, but I was wrong. I urged you to go too soon, and your life became dominated by fear. Instead of seeing the beauty of the experience, it became anything but that. I don't want David to have the same thing happen. He shouldn't go until he is ready.

"Yes, David," Philip said, somewhat surprised to see his hand raised. "Go ahead."

"Well," David said, and his eyes were darting around the room, and everyone could feel his uneasiness. "Of course I want to go, but…" His voice trailed off.

Philip said nothing, waiting for him to continue,

"But you're right; it is fear," David said as his eyes darted around the room, finally coming to rest on the desk in front of him. After a moment, he looked up at Philip. "OK, I'll do it, but you'll have to help me," he added hurriedly, as if he might change his mind. "Tell me what I need to do to get ready." His voice trailed off in a combination of relief and a reluctant resignation. He was glad he had finally spit it out. He felt he was among friends who only wanted the best for him, and that made him feel, if not comfortable, then at least a little safer.

Chapter 8
Fear and Guilt
Or Why You Drag That Anchor Around

"I'll be glad to David; didn't you just know it?" Philip offered with a hint of humor in his voice. The whole class laughed as the tension of the past few minutes evaporated.

"First of all, David, the most important thing to remember, as I said earlier, is that 97 percent of everything a person in Earth Class worries about never comes to pass anyway. So all that fear and worry is, for the most part, nothing, but its impact certainly can't be considered nothing. Fear can literally paralyze your body, and that's just the beginning of its effects. The important thing to remember here, David, is that your physical existence in Earth Class will be a carbon copy of your expectations. What I mean to say is, you will create the exact experience mentally before you project it. If that experience is filled with fear and anxiety, it will be because fear and anxiety were what you believed in and hence what you projected. Remember, first you project your experience, then you see it, and because you see it, you believe it, without ever realizing that you were the one who projected it in the first place. Don't you see, David, the physical experience that you are so afraid of is in your control. In fact, it is your creation, nothing more, but certainly nothing less." Philip paused, waiting to see what effect his comments would have on David.

"Yes, but based on what I've heard, I could forget that I'm the one who's creating it all, and things might go very badly for me. What would I do then? I mean if I forget?

"David, you won't really forget. I mean truly forget. It will always be there, even if it is just a bit under the surface. You'll just have to spend a little time and some energy getting in touch with what you already know," Philip said with genuine concern in his voice. He knew David was on the edge now and could go either way. He wanted to reassure him, but at the same time, only give him the truth. It was, after all, going to be David's decision regardless of what Philip hoped he would do. David had to be assured in his own mind that if he made the decision to go, he was going to be able to succeed.

"How will I know that?" David almost pleaded

"Your comfort, David. Your comfort will tell you. If things are progressing pretty much as you want them to, then you will, for the most part, feel comfortable. On the other hand, if your existence is filled with anxiety and, yes, fear, then you will be quite uncomfortable and there will be an impetus to change. Listen to your inner self, David; it will help you get back on track. Remember, you only fell off the track because you didn't listen to yourself. Somewhere along the line you got a little too caught up in the movie. You forgot that you were in charge. But don't worry, no harm done. Just get back on track. Get quiet, listen to yourself, and just let yourself remember. You may find it a little difficult at first, but your answers and your direction will come to you. The time it will take is only limited by the intensity and the degree to which you are invested in the old beliefs that got you off track in the first place. As always, David, it's up to you. It's always up to you."

"You keep saying remember. Remember what? How can I remember something I never knew in the first place? How could I remember? How will it come to me?" David said with a tone of urgency.

"Start by remembering what you have learned here in class today, for one thing," Philip replied. "Now I suppose you are going to ask me how you will be able to do that, right? Well consider this, David; when something comes into your mind that you have never thought about before, where do you think it comes from?"

David shrugged, shaking his head slowly from side to side, indicating he didn't know.

"It comes from the places you've been to before. You may not have a conscious memory of it, in fact, you probably won't, but locked inside your subconscious mind will be a memory of everything you've learned since the beginning of what you will come to call 'time.' Don't worry that it might not be there or that it will fail you; it will be there, it won't fail you. All you have to do is let it come to the forefront. Let it surface to the level of your conscious mind, where you can use it and act on it. It will come if you just quiet yourself and give in to yourself and *remember*."

That was a lot for David to think about, and he sat silently, letting it all sink in. As was his style, Philip waited for a moment for just that purpose and then resumed, on a new tack.

"All right then, class, and this is not just for David. This next part is a little complicated." Philip seemed to be searching for just the right words, as if what he was about to convey had special meaning. "Things in Earth Class are, as you know, not always what they seem. In fact, it would be more accurate to say that they are frequently just the opposite of what they seem. For example, you may wonder why some of those bad things that David is concerned about do, in fact, happen, especially if you are supposed to be in charge of everything that goes on or happens. While the answer is quite simple, it may sound a little confusing at first. But listen to this. Bill, you confirmed that some of the things that David is worried about do, in fact, happen. Have you ever wondered why they happened?"

"Boy, have I!" Bill responded, sitting up in his chair and leaning his forearms on the desk. "I'm glad you asked, because I've wanted to get a few things straight in this area for some time now. Several times in my previous classes I've wondered why, if we have absolute control over our experience, as our instructors continue to drum into our heads, did some of the things happen to me in Earth Class that happened? I mean, I had some pretty unpleasant experiences, and I'm having a bit of a hard time trying to figure out why they happened, since I was in charge and all." Bill finished sarcastically

"That's exactly what I was hoping you would say, Bill."

Oh no, Bill thought, and his eyes rolled up slightly in his head. Don't tell me I'm about to become the butt of yet another example.

Sensing Bill's uneasiness, Philip chuckled and said, "Relax, Bill. I mean you just provided me with the perfect lead-in I was looking for. You see, class, when I said nothing happens to you that you don't want, I was speaking the truth. I know it sometimes seems like things happen to you that you did not command, but I assure you that is not the case. It's just that, and this is the complicated part, everything that happens is really defined, not by itself, but by its opposite." Philip hesitated again and looked around the room, meeting mostly blank stares. It was clear that no one understood what he had just said.

"Let me put it this way," he continued. "A thing, any thing, is never truly defined by itself, but rather derives its meaning from its opposite. For example, hot would have no meaning for you in a physical form, unless you first knew what cold was. Up would mean nothing to you without down to gauge it against. Left would never tell you in which direction you were going if it wasn't for right. Good would have no meaning if there were no bad. You see, class, the universe *really is* unfolding as it should." Philip chuckled at his own humor.

"To understand a thing, you must also understand its opposite. If you never experienced its opposite, you would only have a partial understanding of that which you desired."

Sally suddenly sprang to life, her eyes wide open exploring Philip's every gesture, and said, "You mean there has to be evil in the world and evil people just so we can see and know the difference between evil and good?"

Philip just stood with shrugged shoulders, his outstretched palms facing up, and raised his eyebrows as he said, "It would seem so, Sally."

Sally was quiet. She had endured a lot of assaults on her belief system since she'd been here, and she was now a little hesitant to jump right into the middle of this one.

"Does that make any sense to you, Sally?" Philip asked.

"A few days ago, I could have said with great assurance no, but now…I guess in a weird sort of a way I can see how that would be possible. It's just that it goes against everything that I believed in when I was…I mean in my last experience in Earth Class. So, does that mean that the police need the crooks?" Sally went on, confident that Philip would take exception to her obvious over-exaggeration.

"I think that's exactly what that means, Sally," Philip said, smiling. Knowing that Sally was not in full agreement, but choosing to ignore it, Philip went on.

"When the cast of a Broadway play finishes its performance for the evening and everyone wants to go out for a cup of coffee, do you suppose the actor who is playing the part of the villain is not welcome to come just because he is playing the part of the bad guy? Of course not! The cast realizes that he is necessary to make the play successful. If it is a play about good triumphing over evil, then evil has a part to play, and the actor who plays the part is serving not only the rest of the cast, but the audience as well. And so it is with life in Earth Class. Some actors will play the part of evil so as to serve those of you who may be striving for good. Just remember, when your experience in Earth Class is over, be sure to include those actors when you go for coffee; they are your brothers, and they may have sacrificed much just so you could play the part you wished to play. So yes, Sally, the police do need the crooks."

Chapter 8
Fear and Guilt
Or Why You Drag That Anchor Around

No one in the class had ever looked at it in quite that way before, and they were digesting what was just said when Philip started speaking again.

"Of course, ultimately it all boils down to a matter of perception. For example, the elephant always looks different when seen from afar. When you are down close to the ground holding on to the elephant's foot, you only get a perspective of the elephant from that angle, which is quite different from the view while holding on to his trunk or tusk, and certainly the view is vastly different when holding his tail. Yet when you back up and see the elephant from a distance you can clearly see that all the parts of the elephant are just that—parts of the elephant. The true picture of the elephant can only be appreciated when you see it in its entirety. So it is with life in Earth Class. Things that you may fear and want to reject from one perspective when seen in their entirety, will reveal themselves to be all part of the same thing, neither good nor bad, and you will see that fear is an inappropriate response to them. Unreasonable fear won't help you with anything, and all fear is unreasonable. Its reality can only be judged by you, and that only by the depth of your belief in it. It is not real! A cursory look around at your brothers will confirm this. Many of them will fear things that others will give no attention to at all. If something were to be feared, all should fear it and not just some."

"Well, some things are feared by everyone," Wayne interjected.

"Really, Wayne? For example?" Philip challenged.

"Well, for example, going to war," Wayne volunteered.

"Do you really think so, Wayne?" Philip questioned. "I don't think everyone fears war. Some won't have to get anywhere near a battlefield, and others will actually profit by it, and there are many other examples of people involved in a war who are not afraid of it at all."

"OK, maybe not war itself, but the fear of dying. Everyone is afraid to die. I'm pretty sure that everyone fears that," Wayne said, looking around the room for support.

"What do you think about that, class? Does everyone share that same fear? What about you, Bill, are you afraid of dying?"

Bill thought for a moment before replying, "I think at one time I was, but I don't think I ever will be again."

"Interesting," mused Philip. "And why is that?"

"Well, because I recently went through the experience," Bill said with a slight smile on his face. Then glancing around the room, he quickly added, "as we all did. And I know that it isn't really anything to be afraid of. Heck, when I first met Tim, he was on my porch, and I had already died, but didn't even know it then. It wasn't a bad experience; in fact, I even had a choice if I wanted to stay or go. I'll never fear it again. At least, I don't think I will." He looked somewhat reservedly up at Philip, who just stood motionlessly with no expression.

"I wonder if I will know about that when I get to Earth Class next time, or if I will forget," Bill continued, his voice trailing off in thought. Philip just stood silently with a "who knows?" look on his face.

Chapter 8
Fear and Guilt
Or Why You Drag That Anchor Around

"Is that what you mean by remembering, Philip? Will I remember and not be afraid the next time. Is that possible?" Bill questioned.

"Not only possible, but it happens all the time," Philip replied.

"You mean that there are people in Earth Class at any given time who actually do not fear death from a conscious point of view?" Bill asked.

"That's what I mean, Bill," Philip said matter-of-factly. "So you see, class," he said, looking up at the class, "even dying is not feared by everyone. Oh, I know it is a fear that is shared by a great many people, I won't deny that, but it is not absolute. You will notice the people who fear it the most are frequently the most unsettled in many other aspects of their lives. They have not been able to give themselves the time they need to get quiet and go within. If they did, they wouldn't be so unsettled in the first place. Many of the fears that they have are nothing more than a habit, unfortunately a bad habit that doesn't serve them at all. Therefore, class, I submit that *fear is an inaccurate conclusion based on a false conception, founded in nothing, based on nothing.* One who believes too much in the opposite, forgets the purpose that the opposite serves and acquires fear. It is simply a decision, a choice that is made, nothing more. It has no reality except that which its author gives it. It will hold you back more than anything else you will ever encounter. You would do well to ask yourself this question when confronted with a 'fearful' situation: *what would I do if I wasn't afraid?"*

Chapter Nine

Acceptance
The Ultimate Key to Happiness

"Good afternoon, class," Tim said with the same smile on his face to which most of the students had by now grown accustomed.

"I see some familiar faces here today. Apparently we haven't bored you all too much the last few days."

The class laughed. Boredom was the last thing they had to worry about.

"Happiness," Tim said, looking up at the ceiling as he searched for just the right words. Shoving his hands deep into his pockets and walking slowly across the front of the room. "That elusive state of mind. That condition which you will spend so much of your time in Earth Class searching for. Those of you who have already been there can attest to the others to the validity of what I am saying."

Several members of the class looked at one another and nodded in agreement.

"Yet why, you might ask, should it be so elusive in Earth Class, when it is the natural condition everywhere else—well, almost everywhere else," Tim added as an afterthought. "Why do you suppose this normal state of being is pursued so diligently and with such little return for most of its pursuers? Anyone care to venture a guess?" he said, looking over the class.

"Yes, Wayne," Tim said, acknowledging Wayne's raised hand.

"Well, I can tell you why I was so unhappy," Wayne said, looking around the room. "As most of you know, I only recently graduated from Earth Class myself, and the memory of it is still quite vivid. I was unhappy most of my life. My childhood was terrible. My mother was a very angry person and was always beating me up. I was scared almost every day of my life. When I reached adulthood, things weren't much better. I was sick most of my life, in and out of hospitals. I couldn't get a good job, not only because I was sick, but also because I wasn't very bright. Only low-paying jobs were available to me, which in turn made things even worse, because I never had very much money. So yes, I think I know a little about the poor return on the happiness investment. Who could be happy with the set of conditions that I ended up with?" Wayne glanced at Tim. Somehow he just knew he wasn't going to get away with that.

"Listen," Wayne said, looking directly at Tim. "I know you are going to say that I chose all that, but that's hard for me to swallow. I mean, why would I choose something like that?"

"Indeed, why?" Tim said, focusing his attention on Wayne.

"Oh boy," Wayne said under his breath. "Here we go again."

"I'm not going to put you on the spot here, Wayne, and seriously, I understand this can seem very confusing, but then isn't that why we're here? Class, the problem is really very simple, and I'm not trying to patronize you. Follow me along here. The reason that happiness seems to avoid you is because you have not accepted your position."

"Gee, Tim," Sally injected. "Are you saying that Wayne should just accept all that stuff that happened to him and not try to better his position? Isn't that a little contrary to what you and some of the other instructors have been saying?"

"Excellent question, Sally. And the answer is no…and yes. No, in that a very important part of the physical experience is to learn how to improve yourself, and to that end, effort should be made. But don't confuse improvement with happiness. Look at it this way. You have essentially two choices. You can *do what you are* or *you can become what you do*. One is very easy and one is very hard. Anyone care to venture a guess which one's the easy one?" Tim looked around the room for a volunteer. No one spoke up.

"All right then, *doing what you are* is the answer. Anyone who was going to say that can now take a bow," Tim announced with his usual sense of humor. "Why?" he asked rhetorically. Without pausing for an answer, he continued. "Because you are already doing what you are. Everyone does what he is. If you're a painter, you paint. If you're a thinker, you think. If you're a thief, you steal. Well, you get the point. It's not difficult to be what you are. The rub comes when you try to *do* something to *become* something. That method will work somewhat, but it requires a lot of effort, and the first lack of attention will cause it to erode if not outright crumble. Therefore, it's much easier to *do what you are.* Now for the second part of that question, Sally. Yes, you should always try to improve your condition. What I have just been talking about is how to go about it. But remember, we are talking about happiness here. How to be happy! You don't get happy by doing things. You just *be* happy." The class was just sitting as if waiting for the other shoe to drop. "To be happy, you must first answer this question: Do I like what I want? Or do I only want what I like? Well, I see heads are starting to spin again. OK, let's break it down."

"Bill, do you ever want things that you don't like?"

"Ah, I can't imagine that I do, no." Bill said matter-of-factly.

"Oh, is that so?," Tim asked with a childish tone. Everyone in the class knew this was going to be good. Tim loved to good-naturedly set the students up like this. "All right then, when you were in Earth Class last time, Bill, did you ever get a bad cut that you had to go to the emergency room to get stitches?"

After a moments hesitation, Bill said, "yes, I'm sure I did. Three times I believe"

"And do you remember getting a tetanus shot?"

"Yes, I remember getting at least one."

"Did you like it?"

"No, of course not, they're painful."

"Well then," Tim said and hesitated. "Why did you get one if you didn't like it?"

"Because if I didn't I could end up in a worse condition."

"So you did want it then? You didn't like it, but you still wanted it. Is that correct?"

"Yes, I…I guess so."

"Well, be sure now, Bill."

"Yes, I'm sure. I wanted it"

"OK, then, so you did want something that you didn't like. Is that what I understand you to say?"

"Yes, I guess it is."

" Hmmm, interesting," Tim mused. "Wait a minute," he said. "Maybe it wasn't the shot you wanted when you went to the hospital. Maybe it was the cut you wanted. Do you think that could be it, Bill?"

"No, Tim, I don't think so," Bill said smilingly. "I can assure you I didn't want that cut."

"Hmmm," Tim mused again. "Wasn't your wife a nurse, Bill?"

"Yes, She was."

"Where did you meet her, Bill?"

"I met her at the…" Bill's voice trailed off.

Without waiting for him to continue, Tim interjected, "In the emergency room?"

"Yes," Bill answered, sitting back in his chair lost in thought, his face devoid of any expression.

"Wow," Wayne said. "That's creepy."

"Not really, Wayne. You see, class, every action has consequences. They're not necessarily bad; in fact, most aren't. Everything you think, say, or do produces a new action, which is just a consequence of an old action. It's just that sometimes they're hard to string together to see that one action has something to do with another seemingly unrelated act or consequence. Anyway, you see now that Bill's cut was an action that for him produced several consequences, or think of them as new actions. Not the least of which was meeting his wife to be.

"OK, let's take another tack here. Let's say that you have had trouble being happy. You've finally stopped being a victim and realized that you are the one in charge of your happiness, and no one else can take it from you unless you give him or her the power to do so. Now you are going to chart your own course. You are going to make yourself happy. 'OK,' you say, 'let's see, what shall I do?' After a few minutes of deliberation, it comes to you. You will only do the things that you like. That will surely bring you happiness."

All eyes were on Tim. This was some good stuff, and everyone felt the secret to happiness was just around the corner.

"So you start to make a list, a list of all the things you like. Because you are a positive person, in just a short time the list grows quite big. 'Let me see,' you say, 'I think I'll pare this down to just the top twenty. After all, twenty of my favorite things should keep me occupied for a while.' You realize that you can always add more later.

"So, in a few minutes there you are, you and a Top Twenty list of your favorite things. Surely, happiness can't be far off now. It's not long before, human nature being what it is, you realize that there are some of the things on your list that you like better than others. So you do what any intelligent person would do. You prioritize your list. You put your *favorite* of the favorites on top, descending to the least favorite of your favorites at the bottom. 'There, that's better!' you exclaim. Now you can focus on only your most favorites. Now let's say, for the purpose of our little exercise here, that your most favorite of the favorites is…ah…Eating Chocolate! OK, now what happens? Well, naturally you focus on the top of your list, and it isn't long before the twentieth item on your list is somehow relegated to the 'things you don't like list,' when only moments before you were sure it was something you liked, and it deserved a place on the most favorite list. Well, I know you can see where I am going here, class. Soon you find yourself sitting on the floor in a chocolate factory stuffing yourself full of chocolate day in and day out, because, after all, it is your favorite! Sounds like fun, huh?"

Everyone looked at each other with an expression on their faces as if they were about to throw up.

"What's the matter with that?" Tim said with a puzzled look on his face. "Hey, you are doing your *favorite* thing, aren't you happy?"

"No. I, for one, sure wouldn't be," Bill piped up

"Why, Bill? Don't you like chocolate? Did I pick the wrong example?" Tim asked, again with a puzzled look.

"No, Tim, it's not that. I think you know what it is. A person needs more than just one experience to make him happy."

"Thank you, Bill," Tim said. "Of course he does. By the way, Bill, how many does he need?"

"Ah…I don't know. I don't think it's any particular number. Just enough to fulfill himself and make him happy."

"But you think there is a number, some number, right?"

Bill didn't answer. Tim was just too sure of himself, and Bill could feel him boring in on the objective. No matter what Bill said now, Tim was going to use it to make his point.

Sensing Bill's reluctance, Tim went on, answering his own question.

"Class, everyone who has had an experience in Earth Class knows there is a natural tendency to gravitate to the things that one likes. As mentioned earlier, feelings of like and dislike play important parts in pointing you in the general direction that you have decided you want to progress. That said, however, your happiness is never dependent on anything you do, but rather on how you feel about what you are doing. Some people like to climb mountains, while others wouldn't consider it. Some people find happiness in knitting a sweater; others would find that tantamount to torture. Some people can even find a modicum of happiness in a prison cell, while others would literally claw the walls to get out. Class, it is never about what you are doing, it is always about how you feel about what you are doing. No amount of doing will accomplish anything toward finding happiness if your only reason for doing a thing is in the false hope that it has the power to make you happy. Life is not an Easter egg hunt. You do not just keep turning over rocks until you find happiness. The key to happiness is learning to like what you want, not wanting only what you like. And, class," Tim said, hesitating until he was sure he had everyone's complete attention. "You *do* want everything that happens to you! If Bill hadn't cut himself he would never have met the love of his life. I'm sure you would never trade the cut that took you to that emergency room for anything, would you, Bill?"

Bill only smiled, nodding slowly in agreement, his mind drifting back momentarily to the memory of his wife of sixty-one years.

Tim went on, "You see, class, as you have heard many times, you have free choice in deciding what you want to pursue in Earth Class. You will generally set the conditions for yourself before you arrive in Earth Class, but you are free within those general conditions to experience whatever you like/want. The words 'like' and 'want' are really interchangeable. It is only your false perception that something could happen to you that you did not want that makes you give a different definition to the word 'like.' What you will need to do is learn to accept that everything that happens to you is something you want, and it is serving the end that you created it for. Happiness will find you when you finally realize that everything that happens to you is, on one level or another, something that serves you in your overall objectives of an Earth Class experience. Understanding this will make it a lot easier to learn to like the things you want and shake off the limitations of your thinking that make you believe you only want the things you like. Wow," Tim said, laughing at himself. "That was quite a mouthful, but do you understand what I am saying here?"

Most of the class nodded a somewhat reserved yes to Tim's question.

"I hope so, because not liking a thing because you believe you don't want it will cut short your view of the reason why you attracted it to yourself in the first place, thereby reducing the experience from all that it could have been.

"Well, I think that's about all for the day, class, unless there are any questions," Tim said, walking back to his desk and putting some papers in his briefcase. It always made Bill chuckle to himself how all the instructors used real props in the classroom to make the experience seem more vivid to the students. Bill knew there was nothing on the papers that Tim had shuffled and put in his briefcase. Even the briefcase itself wasn't necessary.

"You're catching on, Bill." Tim said as he looked up and winked at Bill. "Looks like you'll be getting to head back soon, huh?"

Bill was shocked that Tim might have read his thoughts, but then, why should he be, he thought. Words are just props too. The two men shared a quick smile that most of the class missed as they were getting up from their seats and heading for the door. Bill stayed behind for just a moment. He wanted to say something to Tim.

"Tim," Bill said, then hesitated as though searching for just the right words. "I just wanted to thank you for all you've done for me. And the others, of course," he added quickly

Tim interrupted, "Think nothing of it, Bill. I'm glad to help." The two men looked into each others eyes, not saying a word, yet a feeling of gratitude was given and accepted that transcended words.

Chapter Ten

Ritual, Religion, and Superstition
Or What Other People Think You Should Think

"Good afternoon, class, my name is Ariel. An unusual name for many of you, I know. It means 'high place.' I guess my parents must have had high hopes for me. My sister tells me that it just means that I should move upstairs to the attic and let her have my room."

The class chuckled softly, now accustomed to the humor many of the instructors were so comfortable with.

"All right, let's get to it, shall we?" Looking over the class, Ariel took in a slow, deep breath, stood up straight and started to speak.

"Many of you will soon be embarking on your next Earth Class experience. For you, this is a busy time. Consideration of upcoming conditions and events that you want to experience in this next Earth Class are paramount in your mind, and so they should be. To a large degree, your Earth Class experience will be determined by your decisions of the next few days. Will you be born rich and beautiful, or will poor and disfigured serve your objectives better? Will you be intelligent or dim, black or white? Which nation will you select, and what occasions in upcoming world events will serve you best? While you will always be free to change your conditions within the limitations that you will set before leaving here, the initial conditions that you select will go far in providing a fertile ground for your Earth Class experiences.

"Shortly after your arrival in Earth Class, your 'education' will begin. Part of that education is learning what others before you have learned. You will use your own filters to determine which of the many opinions will serve you and your brothers best. Some of your instructors will only mildly influence you, and others will have a great impact. Still others will try to have a great impact and want you to accept what they believe and what they believe only. Their passion for their beliefs can be quite intense. In fact, their passion can be so intense that it can literally blind them to any other considerations. These people have frequently derailed themselves from their own objectives. They have accepted other people's objectives as their own, and they will try to make you accept them as well. They believe the more people they can convince, the more valid their belief must be, but remember, class, even if it's 50,000 people who have the bad idea, *it's still a bad idea!* Fortunately, these people are easy to spot. People who don't know what to do for themselves are usually the loudest in trying to tell you what to do. Trust yourself in these matters. Get quiet and go within. You will find that you already have everything inside you to decide what it is that you need to do. Many people are deceived by appearances. *Dare to think for yourself!*

"All right, you get my point. Any questions so far?"

Sally raised her hand, surreptitiously looking at the other class members, to see if anyone else was having the same thought, even before she spoke.

"Sir," she said, her voice halting and uncertain. "I can't help but think that you are referring to religion when you say to be careful before accepting what others have to tell you. My question is, how can we know for ourselves about every matter that's in the Bible? Don't we have to accept things like that on faith?"

"Sally, you are free to accept what you want to accept, whether it is on faith or not. I am only telling you to choose what you want to accept. Don't let it be forced on you by others. Many people are afraid, and they accept things others tell them out of fear that their own opinion is somehow of less value. I assure you, it is not. Accept what you want always, but do it for your own reasons, not someone else's. This isn't always easy. In fact, it usually isn't; but it's always the best way. Accepting blindly what other people preach has led the world down some of its darkest paths, not the least of which was the Inquisition and the Dark Ages themselves. The Age of Enlightenment was born out of the thoughts and actions of freethinking individuals, one person at a time. Let me tell you a little story about what happened to me once. OK?"

"Of course," Sally replied. "Please."

Ariel walked around to the front of his desk, folding his arms across his chest as he leaned back against the desk and started to speak.

"Like many of you, I have also experienced many lives in Earth Class. During those periods, I learned a great many things, things that many of you are only now coming to realize. I found one of the great secrets that we all are looking for. When I realized it, I was overjoyed and couldn't wait to share what I had learned with everyone. The message was simple and beautiful, and it provided the answer to all of our questions. When I started to explain my experience and share what I had learned, purely for the benefit of others, I found that many people, many just like you, were a little hesitant, even a little resistant to learn something new. It was, for them, much easier to just 'go with the flow.' The new teachings, though they would certainly lead to their liberation and a sense of joy, peace, abundance, and love far beyond anything they had ever experienced, just seemed too risky. It did require effort, but that was not the problem. It required that they accept responsibility for themselves and stop seeing themselves as victims. But it was tough for them, just as it may be tough for some of you. After all, being a victim is much easier than being a perpetrator. As a victim, it is very easy to see the things that are happening to you as something somehow outside of your control. Something you can do very little, if anything, about. This brought them a sense of comfort. After all, if they couldn't do anything about it, then there was nothing to do but accept and endure. Accepting and enduring has very little to do with what I was trying to teach; that each person is the master of his own destiny and that each has to accept responsibility for himself. Eventually, though some listened, most grew tired of my teachings. They were disturbed, however, at my sense of calm and peace at a time when few possessed such qualities. Their discomfort led to anger when I would not renounce these strange new beliefs—that we were all free to do what we wished, and we needed to set aside anything that tries to remove us from this most natural state, whether it be religion, ritual, superstition, or anything else. Well, you can imagine that last comment did not endear me to the prevailing religious community. In fact, they gave me an ultimatum; either recant my teachings or suffer the most severe consequences. Of course, I realized that there could be no more severe a consequence than to do just that. Eventually, they had enough of me and they removed my physical body from their midst. They thought they had killed me and would be finished with me, but the Truth can never be extinguished. They only hurt themselves by not learning what I had learned and enjoying the happiness that I enjoyed during those days and continue to enjoy even now. Class, I say to you now, what you see and what you experience is an exact result of what, for one reason or another, you want to see and want to experience. Learn this truth and you will never again be a victim. Never again will you suffer because of something foisted on you by others. You will simply change your desire, and your experience will change to accommodate it. I know of what I speak. Follow me, my brothers, and this will set you free."

Just as Ariel finished speaking, a glow enveloped his body and the entire room took on the same glow. The class was dumbfounded and no one spoke. Wayne slowly rose, and there were tears streaming down his otherwise emotionless face. He was sure now who this man who called himself Ariel really was. He walked in a slow but deliberate manner to the front of the room, stood before him, then dropped to his knees saying,

"You are the Christ, aren't you?"

Wayne was completely overcome by emotion. Sally's hands immediately went to her cheeks. Her mouth open in awe, she stared at Wayne kissing the feet of the instructor.

Ariel reached down and helped Wayne to his feet.

"Rise, my brother," He said. "I do not need your worship. I am complete without it. I have no needs at all, as you do not. The only difference between us now is that you do not realize that. I appreciate your respect and love, but you have been deceived. I am without wants. I do not need your worship. This I say to you: There is nothing I do that you cannot. Did I not tell all of those who listened to me in Earth Class that what I do, they could do and even greater things? As long as you believe that you cannot do the things that I do, then you cannot. Walk with me now, and I will show you how wonderfully wrong you have been."

Ariel put his arm around Wayne, steadying him as the two men stood at the desk. He smiled gently at Sally.

"Yes," he replied at last to Wayne's question. "I am the Christ, but so are you, just in an unrealized form. Do you think the Father only has one child? We are all Sons of God. I have the realization of that fact residing deep within me. That is the *only*, I say again, the *only* difference between us now. The experience that you are all about to embark on is just another step in the realization of who you really are. The 'time' it takes for each of you to realize it is up to you. Only the outcome is certain…but all of this is for another time and another class."

The class gathered around Ariel, each person pressing against the other in the circle, quietly, no one speaking a word, each feeling the incredible love emanating from Ariel and, in turn, each other.

Bill was quiet as he left the room, as were the rest of the class. Gone was the usual bustle of conversation about the just-completed class. Each person was alone with his own thoughts and yet at the same time feeling very connected to one other. It was a feeling of euphoria, and Bill was savoring it.

Chapter Eleven

The Beginning
Or What Are All Those Bright Lights?

"Good morning, ," Tim said as he entered his office. Bill had been waiting for about five minutes when Tim walked in. Tim had left a note for Bill in his room the night before saying that he wanted to see him in his office in the morning. Bill wondered what this was all about, as Tim had never seemed so official before. He couldn't help but feel a little like a grade-schooler who has just been sent to the principal's office.

"Good morning, Tim," Bill said somewhat apprehensively. "What's up?"

"Nothing, really. I just wanted to say good-bye to you before you left for Earth Class," Tim said.

"Well, that's nice of you, Tim, but I don't think I'll be going any time real soon," Bill replied.

"Really? Hmmm, I thought you were leaving this evening," Tim said as he shuffled through a small stack of papers. "Yes, here it is: Andrew Thornton, born to Martin and Elisabeth Thornton on June 2. Looks like you'll be leaving a little sooner than you thought."

"What? What are talking about? Who are these Thorntons? What are you talking about, Tim?"

Tim looked at Bill with raised eyebrows and puckered lips that conveyed a "what do you know about that?" look.

All of a sudden, Bill's eyes opened wide and his jaw dropped.

"You mean…!" Bill exclaimed, not finishing his sentence.

"I think so, Bill. Congratulations!" Tim stood up from behind his desk and offered his hand to Bill.

Bill accepted Tim's hand halfheartedly, still in a bit of a daze.

Tim, sensing Bill's uneasiness, spoke up. "It's been my pleasure knowing you and helping you out while you have been here. I was honored when you selected me to help you 'graduate' from your last Earth Class experience, though it was nip and tuck there for a while on your front porch. We've been friends for a long time, Bill, and you are going to know the whole story soon. In fact, if all goes well, this could be your last Earth Class cycle before going on to an instructor's position."

"What, me an instructor? I don't think so!" Bill exclaimed. This was just the latest in a list of surprises. "I've barely got a grasp on all the stuff I've been exposed to here. I certainly don't understand it enough to teach it, and what to you mean my last Earth Class cycle?"

Chapter Eleven
The Beginning
Or What Are All Those Bright Lights?

"Whoa, Whoa! One question at a time," Tim said, chuckling. "As far as you becoming a teacher is concerned, you'll soon see that you're a lot more ready than you think."

"What do you mean?" Bill asked, his eyes searching Tim's face for some sort of explanation.

"One thing that we didn't get into in your classes was any discussion of what comes next after you've finished all your Earth Class experiences. I don't want to get into it very deeply now, because you still have one more Earth Class left, and you have much left to accomplish while you're there. But I guess I do owe you some explanation. I said that we have been friends for a long time, and that's true. You and I had a relationship very similar to the one you and Andy just finished. For now, let it suffice to say that I wouldn't be here today if it wasn't for you."

This last comment stunned Bill. He had always admired Tim. He had no idea that they had had any previous relationship, much less that he had ever helped Tim in any way, especially to that degree.

"So you see, when you made the appointment with me on your front porch, there was no way I wasn't going to be there for you, even if you were a grumpy old fart." Tim's ever-present smile only hinted at the genuine affection he had for Bill.

"Surely you've thought about what comes next, Bill. You didn't just think that you would be stuck in Earth Class forever, did you?" Without waiting for an answer, Tim went on. "Your Earth Class experiences are like pages in a book. No one page can convey the entire story of the whole book. Some pages reveal a lot, some very little, but together they piece together the whole story, and so it is with your life. When you complete all of your experiences, you *graduate* again, only this time it will be a bit different. This time it will be kind of like getting your PhD, roughly speaking, of course. Anyway, many of us who are so motivated, go on to be instructors to help the others in their Earth Class experiences, but again, as always, this will be your choice. When you complete your last experience, just like finishing a book, you will know the whole story. All the parts will come together. You'll realize then that you do know a lot more than you are aware of right now. When you piece all of your experiences together, you'll see a clear picture of why everything happened the way it did. It will be easy for you to make your decision then, and I don't want to influence you too much, but I've seen you in action and you're a great teacher."

Bill just stood still, saying nothing, trying to digest all that he had just heard.

"Now, as far as your second question goes, the Thorntons are also finishing what most likely will be their last Earth Class experience. You and they have a special long-term relationship. I don't want to say too much about that right now, but you can be sure, since there is no such thing as coincidence, that the three of you have some unfinished business together.

"I know you think that this is a little soon and that you're not ready, but I assure you, you are. I've discussed this with the other instructors, and they all agree; you're ready for this next Earth Class. The parents you're getting are fantastic. As I said they are about ready to permanently graduate from Earth Class themselves. I know we haven't talked about this at all, but Martin and Elizabeth are very old souls, and they are soon to complete their Earth Class cycles and move on to that higher dimension that we were just talking about. They will be very helpful to you, Bill, and they are eagerly awaiting your birth. You couldn't ask for more loving parents.

"Look, Bill, I don't want to hurry you, but you've had Elizabeth in labor now for more than six hours just waiting for you to make up your mind. There are several people most eager to take that new little baby's body, but she is holding out for you. What are you going to do?"

Bill smiled and shook his head. "Boy, you are just full of surprises," he said, smiling at Tim. The two men embraced, and Tim patted Bill on the back.

"Well, I guess this means I'll see you soon, huh, Bill…ah, I mean Andrew?"

Oh well, he was sure this, like everything else that had happened since he got here, would become clear soon enough. After a short hesitation as the two men looked at each other, Bill gave Tim a smile and, still shaking his head, said, "Yes, I guess so. Thanks, Tim, thanks for everything." There were tears of happiness in his eyes. "I'll never forget you."

"Of course you won't. I'll be seeing you sooner than you think." With that, Tim winked and walked out of the office, leaving Bill wondering just what that comment was going to turn into.

Well, what the heck, Bill said to himself. No point in hanging around here any more, and with the same quick resolve that had always been his trademark, he put his hands up in the air and muttered, "Let's blow this pop stand!"

All of a sudden, in the back of the room, Bill could hear people talking. He turned around to see what all the commotion was about, but he couldn't see a thing, the bright lights of the delivery room blinded his eyes.

"IT'S A BOY!" Someone shouted.

Bill tried to speak, but somehow he could not form any words. Every time he tried to say something, only a faint cry came out of his mouth. He felt something in his nose. Now he was lying on a table and someone was saying, "seven pounds, nine ounces." Again he was picked up and moved, finally coming to rest just as someone said, "Here he is. Here is your son."

He couldn't make much of his surroundings, but he could feel arms cradling him and Elizabeth's soft lips kissing him on the forehead. He felt that he was loved and safe, and it felt good.

Sleep, he thought, now he needed sleep. This was a day whose beginning he could no longer remember. The last thing he heard was Elizabeth's voice saying,

"I love you, Andrew."

Hmmm, he thought, Andrew, huh? I guess that's not so bad.

The Start

One more time…

Acknowledgments

First and foremost I would like to express my sincere gratitude and appreciation to all the thinkers and writes that have been my constant companions throughout my life and have so influenced my life; especially Richard Bach and the immortal Jonathon Livingston Seagull who started me on my quest over 30 years ago.

Also my dear friends Mark and Nancy Kotnour, Jim and Sharrone Ault and Terry and Yvonne Twigg who have put up with my ramblings over the years and have loved me in spite of them. And of course my children who gave me all those golden opportunities to practice my famous "lectures". And of course my dear aunt "Beanie" who certainly doesn't understand me, thinks I'm a little weird, but for some strange reason still talks to me.

A special thanks to Pam Wagner, who had to put up with the whole process and whom, to this day has to listen to forming philosophical musings when I'm sure she would rather be sleeping. *(Can I help it if my greatest insights seem to manifest at night)*

Last but not least, Soumitro Das, my editor. I now know why virtually every author thanks his editor. If the public ever saw a book the way an editor first sees it, movies would be a lot more popular.

About The Author

I was born in Seattle in June of 1944 and except for a ferocious competitiveness that occupied the first 40 years of my life; I have lived, at least to all observers, a fairly ordinary existence. That competitiveness has brought me both my greatest rewards including financial success and my greatest heartaches including the loss of two exceptional women through divorce. It has been an intense life and knowing what I know now, I guess, like most of us, there's not that much that I would have changed. I have four children, two boys and two girls, with the youngest, at last, graduating this year from the University of Washington.

After 58 years in Seattle, I recently moved to Phoenix Arizona and I recommend it though the drivers are a little crazy here, but then it is the Wild West. My interests are flying my airplane, trying constantly to comprehend Quantum Physics, (a big job) talking with people and trying to help them realize how powerful they really are, and now writing.

Though I have many diverse interests, Spirituality is high on that list. Though I am not at all a religious person, I have spent a great deal of my life seeking the answers to what I thought were the five most important questions of life:

Who am I?

Who is my God?

What is my true relationship to Him?

Where will I go and what will happen to me when I die?

What should I really be doing while I'm here?

Initially, I sought those answers through traditional religious avenues, but… well you know how that goes no matter what side of that fence you're on. After a lifetime quest, I'm somewhat satisfied with the answers I've found, though there will ***always*** be room for expansion; and I've written some of those answers subtly into this book.